The Politics of the Final Hundred Years of Humanity (2030-2130)

Ian Cook

# The Politics of the Final Hundred Years of Humanity (2030-2130)

Springer

Ian Cook
School of Business & Governance
Murdoch University
Murdoch, WA, Australia

ISBN 978-981-15-1261-2 ISBN 978-981-15-1259-9 (eBook)
https://doi.org/10.1007/978-981-15-1259-9



This Springer imprint is published by the registered company Springer Nature Singapore Pte Ltd.
The registered company address is: 152 Beach Road, #21-01/04 Gateway East, Singapore 189721, Singapore

*To Eliza, Nicholas, and Jack.*
*Of my many failures as a father, this was the worst.*
*I didn't hear. I didn't understand. I didn't act.*
*I'm sorry.*

# Contents

# Chapter 1
# Introduction

[A] 4 °C world is not the plausible worst-case, it is the expected outcome of the emissions pathway we are currently on. The worst-case version would be if the 4 °C world occurred far sooner than expected. … one of the Royal Society articles, 'When Could Global Warming Reach 4 °C?'… contains this remarkable finding: If we stay near the high emissions pathway, and "If carbon-cycle feedbacks are stronger [than currently modeled], which appears less likely but still credible, then 4 °C warming could be reached by the early 2060s in projections that are consistent with the IPCC's 'likely range.'" (Romm 2015, pp. 137–8)

In 2010, celebrated Australian microbiologist Professor Frank Fenner predicted that overpopulation, environmental destruction and climate change would result in the extinction of human beings within one hundred years (Edwards 2010). Fenner was not a climate scientist. He simply listened to climate scientists and recognised that little would be done to prevent what will be referred to here, for reasons presented below, as catastrophic human-caused environment change.

Fenner has not been alone in warning of the potential for the end of humanity. Torres listed a number of other figures who have made similar predictions: Sir Martin Rees, of Cambridge University's Centre for the Study of Existential Risks wrote "in his 2004 book *Our Final Hour* that there is a 50/50 chance of civilization being destroyed before the 22nd century. … similarly pessimistic figures can be found in the work of John Leslie, Sir Nicholas Stern, and Nick Bostrom, the founder of the Future of Humanity Institute at Oxford University…" (Torres 2016, p. 60). As the title of this book suggests, Professor Fenner's prediction was a reasonable one but that it is likely to be out by around 20 years. Focussing excessive attention on the date at which it begins, however, misses the point that too little will be done to prevent humanity moving into its final hundred years and that, whether it is in 2030 or not, this point will be reached.

Considerable attention in this book will be paid to explaining why too little will be done. But the first, and most important, reason concerns the scale of the changes that would be required to prevent catastrophic human-caused environment change from leading to the final hundred years of humanity. A 2018 report in *National*



I. Cook, *The Politics of the Final Hundred Years of Humanity (2030-2130)*,
https://doi.org/10.1007/978-981-15-1259-9_1

*Geographic* listed just some of the changes required to prevent humanity entering its final hundred years:

> no home, business, or industry heated by gas or oil; no vehicles powered by diesel or gasoline; all coal and gas power plants shuttered; the petrochemical industry converted wholesale to green chemistry; and heavy industry… either using carbon-free energy sources or employing technology to capture $CO_2$ emissions and permanently store it. (Leahy 2018)

And even if "between 0.4 and 2.7 million square miles (1-7 million square kilometers) of land … [is] converted to growing bioenergy crops and up to 3.86 million square miles (10 million square kilometers) of forests [are] added by 2050" this will simply stabilize carbon dioxide emissions. The problem is that the "$CO_2$ emitted in the last hundred years will continue to trap heat in the atmosphere for hundreds of years to come. By 2045 or 2050 there will still be too much $CO_2$ in the atmosphere" (Leahy 2018).

This is not to deny that, even in 2020, two paths remain available with respect to catastrophic human-caused environment change. One is the path of collective action on the parts of people and governments in most, if not all, countries, and all of those in the Global North, to massively reduce the effects of human activity and prevent the catastrophic changes to the environment that are threatening the end of humanity. The other path is that of widespread inaction with limited, *ad hoc* and, at best, piecemeal efforts on the parts of people and governments in some countries to address catastrophic human-caused environment change. This leads to both a profound reduction to the planet's habitability and a deep disturbance of human societies and international politics to the point that humanity ends. This book explains both the politics that has and will continue to ensure that the second path is followed and the politics that will result from following that path.

Before this discussion proceeds, however, it is important to recognize that focusing on global warming and climate change narrows the picture concerning catastrophic human-caused environment change because it directs attention away from two other processes that are crucial to the effects of human activity on the habitability of the planet: species depletion and extinction and ever-increasing toxicity. While neither species depletion and extinction nor accumulating toxins are as immediately threatening as global warming and climate change, both continue to contribute to the decreasing habitability of the planet and the end of humanity.

Despite that fact that it will be repeated regularly throughout this book, the phrase 'catastrophic human-caused environment' will be used in full and not substituted for an acronym (CHEC or CH-CEC – which would probably be pronounced 'check'). This is to ensure that the catastrophic character of the threat represented by human-caused environment change remains at the center of the discussion. And it is to ensure that the 'human-caused' aspect of this environment change continues to be highlighted. And, as noted in the preceding paragraph, to ensure that the focus is on environment, and not just climate, change.

Even though what some call the sixth mass extinction (Kolbert 2014) has not yet reached the scale and scope of the five preceding mass extinction events, species depletion and ultimate loss continues. Wignall has suggested that "current extinction rates may be comparable, suggesting that we may be witnessing the start of a mass extinction episode" (2019, p. 18). Youngsteadt, López-Uribe and Sorenson concurred: "The Earth is experiencing a wave of anthropogenic biodiversity loss, such that current rates of extinction are 100–1,000 times the background rate observed between prior mass extinctions in the fossil record. These losses place Earth's biota in the early stages of an extinction event comparable to those precipitated only five times before in the past 540 million years" (2019, p. 119).

The increasing number and levels of toxins in the environment represent another aspect of catastrophic human caused environment change. For, "more than 1,400,000 chemicals including pesticides, plastics, etc. have been synthesized till date since 1950 and each year 10 million tons of toxic compounds are being dumped into the environment leading to land degradation, soil salinization and contamination of water resources" (Arora 2019, p. 1). The last of these is making access to safe drinking water an increasing problem for growing numbers of people around the world. Arora estimated that around "500 million tons of heavy metals, toxic sludge and hazardous solvents were… released in global water supply in 2007 making it unsafe to consume" (2019, p. 1). Plastic pollution is another growing problem. In the 1970s the annual rate of the production of plastics was "about 50 million metric tons and it has increased to over 348 million metric tons at present" (Arora 2019, p. 1).

Determining the specific regions that will be affected by species depletion and ever-increasing toxicity is difficult, but species depletion and extinction and increasing toxicity remain important factors in the movement toward the near complete uninhabitability of the planet. The principal focus of this book, though, will be on the politics of environment changes associated with global warming and climate change: uninhabitability caused by excessive surface temperatures and humidity, desertification and rising sea levels. These are not the only consequences of catastrophic human-caused environment change. They are simply the most obvious and immediately important.

Even though climate change is amongst the most heavily researched fields of inquiry in contemporary scholarship, few attempts have been made to develop a larger political picture associated with the effects of human activity on the habitability of the plant. In short, limited discussion has occurred, thus far, with respect to the problems that communities and political systems face in responding to the effects of catastrophic human-caused environment change. Many of these problems relate to intra- and inter-national population movements and raise issues of human rights and government responsibilities with respect to such population movement. Others are with respect to the competition over resources that will emerge during the final hundred years of humanity.

The politics of the last hundred years of humanity will manifest differently depending on where people live, of course, but its general patterns are discernible. This book provides only very general patterns that must be applied in considering the challenges faced by specific national and local governments and peoples. Much detail has been sacrificed in trying to capture as many of the dimensions of the final hundred years of humanity that bear on politics, in one way or another.

People and their governments will be challenged in ways they have not previously been challenged. Within nations, tensions will have to be managed that arise due to the combined effects of desertification (including ocean dead zones), rising sea levels, increasing global surface temperatures, species depletion and increasing toxicity (especially that from plastics). In many cases and over time, increasing numbers of governments will fail and many of the people they once governed will seek to migrate to neighboring countries, while others go further looking for refuge. This creates international sources of and points of conflict. International conflict will also occur as governments with the capacity to do so seek to exploit those resources that remain available or become available as polar ice sheets melt and the migration patterns of marine food species change.

But it is not merely a case of internal (national) and external (international) adjustments being required and conflicts being triggered. Another crucial question concerns the way that those in what many, such as George Monbiot (2007), refer to as 'the climate change denial industry' respond to increasing evidence of the catastrophic human-caused environment change. This 'industry' has been crucial to the politics leading to the final hundred years of humanity and it will continue to function during the final hundred years.

Another aspect of the politics of the final hundred years of humanity relates to the 'desperate measures' that may be undertaken by those governments and international bodies with the resources to do something by way of responding to the declining inhabitability of the planet. Radical population reduction is one of the 'desperate measures' that some, probably many, governments will undertake. Geoengineering initiatives by national and international organizations represent the other 'desperate measure' discussed in this book. The focus here (Chap. 7) is on Stratospheric Aerosol Injection programs because these are most likely to be undertaken during the final hundred years of humanity. Undertaking these programs will give rise to a variety of governance problems the solutions to which, or lack thereof, will constitute an important part of the politics of humanity's final hundred years.

The trajectory toward global uninhabitability remains only one possible trajectory – though the window is closing. A prior question, then, relates to why this trajectory can be taken to be inevitable. One set of reasons concern the way that the politics of science has enabled the level of skepticism that presently exists in many countries. By ignoring science's politics, scientists have left themselves vulnerable to the challenges presented by those in the climate change denial industry. Another set of reasons to believe that the final hundred years of humanity is inevitable is that political institutions are *by nature* unable to initiate radical social transformations of the type required to divert from the trajectory toward humanity's final hundred years.

But scientists' hubris and (inevitable) government failure will not be the only reasons that a trajectory toward the final hundred years of humanity will be maintained. What can be thought of as, ideological factors also contribute to maintaining this trajectory – principally, but not exclusively, in advanced industrial nations. The climate change denial industry is important, but there are a variety of other factors that will also contribute, in major and minor ways, to the maintenance of a trajectory toward the last hundred years of humanity. These include: conservatism, the scope of the problem, the scale of the change required, the normalcy of capitalism, selfish individualism, consumerism, a belief in the end of the world on the parts of members of religions (including, and especially, Christians and Muslims) and that many people of the Global South see those of the Global North as responsible for the problem of catastrophic human-caused environment change.

As has been noted, telling a very big story in a relatively short book requires providing only general outlines of the politics that has resulted in a failure to do anything to prevent catastrophic human-caused environment change and the politics that will occur during the final hundred years of humanity. How politics plays out in specific countries and regions will follow the broad principles explained in this book. But there will be local differences, especially in the first 50 years of the final one hundred years of humanity. So, readers will need to adapt the messages of the book to their specific national circumstances.

## Using Near Worst-Case Scenarios

This book is based on the application of the near worst-case scenario with which this book opened. The driver of this scenario is an increase in the average global surface temperature to 4 °C by the mid-2060s (while it was presented there as a worst-case scenario, more dire scenarios can be developed). This creates a variety of effects, of which desertification and rising sea levels are the most important. The climatic effects of this increase in average global surface temperature will not be even, but desertification will cause a significant expansion of drylands and result in a significant decrease in the productivity of many areas of the planet.

This increase in the average global surface temperature will cause some regions to become uninhabitable, as temperatures exceed those that human beings can tolerate. The increase in the average global surface temperature will also cause glaciers and ice sheets to melt, resulting in a 2-m increase in sea levels. The change in temperature will result in an increase in the number and intensity of extreme weather events and, in some instances, storm surges will add another 2 m to the height of the seas affected by those storm surges.

Increasing levels of carbon dioxide in the atmosphere means that more dissolves in the ocean to cause ever-greater levels of ocean acidification. Ocean acidification produces ocean dead zones because acidification prevents the growth of most forms of plankton, corals and shellfish, which prefer an alkaline environment. Whereas shellfish are important, plankton and coral are basic to many of the oceans food chains and the effects of ocean acidification can be devastating. Thus, in this near worst-case scenario, deserts of the sea will spread along with deserts of the land.

While the effects of increasing levels of greenhouse gases in the atmosphere are part of accounts of climate change and global warming, to these effects must be added those of species depletion and extinction and increasing environmental toxicity. A sixth mass extinction event, then, is part of the worst-case. And a crucial background condition is that the health of plants, animals and people will be deteriorating as they are experiencing the effects of global warming and climate change. With the most alarming aspect of increasing toxicity being the release of nuclear materials as rising sea levels inundate nuclear power plants and storage facilities that house the waste from those plants and from the development of nuclear arsenals (Mizokami 2019).

Two reasons justify the adoption of near worst-case scenarios. The first is that it makes sense to adopt the maxim: 'Expect the best; Prepare for the worst.' This is so in general, in politics and in the context of thinking through the politics of catastrophic human-caused environment change. The second reason to adopt near worst-case scenarios is ongoing inaction with respect to addressing the causes of catastrophic human-caused environment change. All indicators suggest that nothing meaningful will be done to reduce the negative effects of human activity on the habitability of the planet and many projections of our environmental future rely on something being done to slow, if not reverse, catastrophic human-caused environment change. Accepting that insufficient numbers of people and governments will take meaningful steps to address catastrophic human-caused environment change leads to the adoption of near worst-case scenarios.

An important point to note in this context is that the Intergovernmental Panel on Climate Change's Representative Concentration Pathways 8.5 is not treated as a near worst-case scenario. This is because it is not a near worst-case scenario. For, as will be discussed in Chap. 2, the political (intergovernmental) character of the IPCC means that its worst-case scenario must be treated as a conservative estimate.

## Why Start from 2030?

2030 was chosen for three reasons. The first is that the failure of the Paris Climate Accord should be obvious by 2030. Representatives of 196 countries signed the Accord (under the United Nations Framework Convention on Climate Change) in 2015. The treaty is due to come into effect in 2020 and the first climate 'stock take' is to be conducted in 2023 and the five yearly reviews that follow this initial stock take will provide ongoing evidence that catastrophic human-caused environment change is progressing to the point that humanity will have entered its final hundred years.

The following are some important features of the Accord and indicate points at which the failure to meet Paris targets becomes glaringly obvious:

> Mitigation (Art. 4) – The Paris Agreement … prescribes that Parties shall communicate their NDCs [Nationally Determined Contributions] every 5 years and provide information necessary for clarity and transparency. …
>
> *Global Stock take* (Art. 14) – A "global stock take", to take place in 2023 and every 5 years thereafter, will assess collective progress toward meeting the purpose of the Agreement in a comprehensive and facilitative manner. Its outcomes will inform Parties in updating and enhancing their actions and support and enhancing international cooperation.

The second reason to choose 2030 is that it is well after the so-called 'pause,' or 'hiatus,' in the rate of increase in the average global temperature that, in some accounts, occurred between 1999 and, possibly, 2015 – though some models suggest the deceleration may continue for a few more years (see Berkeley Earth n.d. and Fischetti (2017)). Records suggest that the rate of increase in the average global surface temperature appeared to slow during that period (with the principal explanation being that the oceans absorbed more heat that many scientists had expected).

This apparent pause undermined predictions of an almost-continuous acceleration in the global average surface temperature. It validated other modeling, however, which predicted both the deceleration that occurred and an acceleration to follow. Insufficient data is currently available (2019) to determine whether acceleration has begun, so it may not be until closer to 2030 that the data will be clear, and we can be more certain that we have reached the point at which we can mark the beginning of the final hundred years of humanity.

The third, and principal, reason that 2030 has been chosen as marking the beginning of the final hundred years of humanity is that a 2018 report from the IPCC identified 2030 as an important moment for determining whether sufficient progress has been made in addressing the causes of catastrophic human-caused environment change (IPCC 2018). As Rhodes wrote, the 2018 report concluded "that humankind has a mere 12 years left, during which time sufficient and dramatic carbon-emission mitigation strategies must be inaugurated to avoid the 'global average temperature' from rising above the 1.5 °C limit which the 2015 Paris Climate Change Agreement aimed for, while pledging to keep it 'well below 2 °C above pre-industrial levels'" (Rhodes 2019, p. 73).

The objective of the IPCC report was to stimulate the sort of radical action that would prevent the scenario upon which this book is based, of course. But the steps required to halve the rate of fossil fuels use by 2030 and then to end their use by 2050 are profound (unimaginably profound). Rhodes made clear the great, indeed insuperable, difficulties associated with avoiding a 1.5 °C increase in the average global surface temperature. In his view, and it is one upon which this book is based, "to meet the 1.5°C target would require economic transformations at an unprecedented rapidity and scale…" (Rhodes 2019, p. 85).

For the global elimination of the necessary 42 Gt of $CO_2$ by 2050 would necessitate, according to Rhodes, "a threefold increase in the implementation of low-carbon electricity… along with a dramatic elimination of essentially all vehicles that use internal-combustion engines, which currently is practically all of them." Meeting this target would also necessitate, in his view, "the development of various zero carbon technologies, and the increasing availability of green finance…" (Rhodes 2019, p. 85). It is something of an understatement, then, for Rhodes to suggest that "the challenges are immense" (2019, p. 85).

## Of Humanity

This book is a discussion of the politics of the last hundred years of 'humanity,' and not the last hundred years of '*homo sapiens*,' 'human beings,' 'human life' or 'humans'. The end of 'humanity' denotes an end of humaneness (especially compassion). Human life might end. But whether it does is not a question for this book. Here the interest is in a process through which people lose our humanity because we are forced into conflicts or to adopt moral positions that represent the end of compassion, care or any of the other attributes of people with humanity. Pockets of humanoid beings may persist on Earth well beyond 2130. But the conditions under which these last survivors live will preclude much, if anything, in the way of humanity.

'Humanity,' then, is here understood as manifesting itself in caring and connecting to other human beings. It involves over-coming tendencies to be self-centered and focusing on our empathic feelings toward others. As Spikins and co-authors argued, "certain emotions, particularly the socio-moral emotions such as empathy, compassion, shame and remorse, appear to be particularly socially significant. Of all of these compassion is perhaps the most evocative of something we feel defines 'humanity'…" (Spikins, Rutherford and Needham 2010, p. 305). More importantly for this discussion, Spikins, Rutherford and Needham identified four "key stages … [in the] the development of 'modern' compassion" (2010, p. 314).

They were not suggesting that early humans were incapable of precursors to "modern compassion." Indeed, they pointed to "archaeological evidence for care of the ill, deep-seated collaborations, sharing of emotions and a capacity to care beyond immediate relationships or timeframes within archaic humans…" (Spikins et al. 2010, p. 317). But they saw "'modern' compassion" as adding to and being different from prior forms of caring. The position taken in this book is that the sorts of regressions associated with surviving catastrophic human-caused environment change militate against the perpetuation of 'modern compassion' (which is treated here as synonymous with 'humanity').

Some people may survive in isolated parts of the world. But even this seems unlikely. But they will breathe low quality air and their environment will be riddled with toxins. And the surface temperature alone will make it hard for them to survive. Not to mention the radiation that will be left over from the nuclear industries (power production and weaponry) – some of which have been inundated as sea levels rise. And the few creatures and plants that would survive the final hundred years of humanity are not likely to be readily edible.

It is possible that those with the necessary financial or political resources may use those resources to enable themselves and their progeny to survive beyond the final hundred years of humanity. A variety of different scenarios might unfold; such as escape to another planet, into space or biodomes. Once again, the humanity of those who choose and who undertake such a path can be disputed. So, the discussion presented in this book remains unaffected by the possibility that some *homo sapiens* will be left somewhere on Earth, another planet or in space and that small groups of them will be around beyond the next hundred years. They will not have humanity in any sense in which the word is used. So, and in short, there may be *homo sapiens* around beyond the next hundred years, but they will not have humanity.

## On Speculation

The most difficult aspect of writing this book is that the analysis developed in this here is based on predictions of the future. I have never written about the future before and would prefer not to do so. But it is necessary here. The book relies on the available science and in this context, and many others that scientists will not admit, none of this science represents Absolute Truth. As will be explained in Chap. 2, science generates very powerful consensuses that function as truth – but should never be mistaken for Absolute Truth. The greatest (indeed only) hope for those of us who are unpersuaded by some sort of climate science conspiracy is that climate scientists might have gotten it wrong.

The possibility remains that the scenario upon which this book is based is overly pessimistic. The book relies on near worst-case scenarios, after all. Perhaps scientists are not taking something into account that would change their calculations about the effects of greenhouse gas emissions, or species depletion and extinction, or toxins in the environment. All good scientists admit that there may be something missing from their calculations that mean that the predictions they make are not accurate. Environmental scientists may be overlooking something that will slow the rate of global warming and ocean acidification or will compensate for the effects of the increasing toxicity of the environment. Perhaps the environment and the creatures in it are more resilient that most scientists believe and will survive the effects of environment change better than they think they will.

Perhaps other scientists are on the verge of a breakthrough that will mean that our governments and companies will be able to introduce technologies that will reverse the effects of greenhouse gas emissions. Perhaps governments or companies will develop genetically modified crops and animals that can survive with little water and at much higher temperatures and tolerate high levels of environmental toxins (and, hopefully, not pass toxins onto the people who eat them). Perhaps negative emissions technologies or carbon dioxide removal will improve to the point that it can be introduced at much more affordable level than this technology currently is. Perhaps nuclear fission will become a reality and be cheap enough so that it can be introduced in time and in enough countries to make a significant difference.

Perhaps significant international action will occur in time to make a major difference. Perhaps world governments could achieve something better than the non-achievements of the 2015 Paris Summit. There are many reasons to doubt that governments and the people who support them will do what is necessary, though, as will be explained in Chaps. 2 and 3 (where an explanation is offered as to why the end of humanity is inevitable). But perhaps they will and human beings across the planet will embark on a quest to save their planet for humanity. In many ways, the only way to ensure the continuation of humanity beyond the next hundred years is for people to embrace their humanity. It is possible. It is just not likely.

The premise upon which this book is based, then, is that nothing will be done to make a real difference by the number of people and their governments necessary. It is more likely that people and their governments will continue to pretend that business-as-usual is OK, even right. In the absence of a lack of international efforts to address environment change we are left with the predictions that are relied upon in this book.

And there is every chance that the predictions upon which this book is based are optimistic. For just as there is a chance that scientists are wrong in ways that make the scenario outlined in this book too pessimistic, there are just as many reason to believe that scientists are wrong in ways that mean that 2030 will mark the beginning of the final 75 or 50 years of humanity. Perhaps scientists are overlooking factors that will mean that the state of the planet will deteriorate much faster than we think it will. Effects like ocean acidification are not as easy to observe as greenhouse gas emissions are. Even though this meant that the way oceans were absorbing carbon dioxide was under-estimated for a while, it also meant that scientists were slower to recognize the extent of the problem of ocean acidification.

Another reason to think that starting the final hundred years of humanity at 2030 might be optimistic is offered by Steffen and co-authors, who projected a "Hothouse Earth pathway". Here a rise in temperature to 2 °C above the preindustrial temperature creates "biogeophysical feedbacks" that "become the dominant processes controlling the system's trajectory" and result in a continued warming pathway. Steffen and colleagues suggested that this trajectory could begin once the average global surface temperature reached 2 °C above pre-industrial levels "because of the risk that a 2 °C warming could activate important tipping elements, raising the temperature further to activate other tipping elements in a domino-like cascade that could take the Earth System to even higher temperatures (Tipping Cascades)" (Steffen et al. 2018, p. 8254). They concluded that "even if the Paris Accord target of a 1.5 °C to 2.0 °C rise in temperature is met, we cannot exclude the risk that a cascade of feedbacks could push the Earth System irreversibly onto a 'Hothouse Earth' pathway" (Steffen et al. 2018, p. 8254).

The fact that the Intergovernmental Panel on Climate Change (IPCC) consistently underestimated and underestimates the degree and pace of climate change will be discussed below. Here note is taken of the following passage from a 2012 article, again in *Scientific American*: "Checking 20 years' worth of projections shows that the Intergovernmental Panel on Climate Change has consistently underestimated the pace and impacts of global warming" (Scherer, 2012).

And it is not just a matter of scientists in the IPCC underestimating the extent and rate of climate change. A 2017 study, for example, found that ecologists and conservation groups had been underestimating the impact of climate change on wildlife. Basically, the authors of the study wrote "that the impact of climate change on mammals and birds in the recent past is currently greatly underappreciated: large numbers of threatened species have already been impacted in at least part of their range." (Pacifici et al. 2017, p. 208).

And there may be some factors that are more difficult to provide for and are harder to rely upon than the problems that are the focus for this book. There can be no certainty, for example, as to the effects of climate wars on the planet. Wars are always bad for the environment. Nuclear wars are even worse. Nor is it possible to know what will happen when leaders who believe in some form of afterlife and whose governments possess powerful weapons, including nuclear weapons, decide that humanity is doomed. Some of these leaders might decide to hasten the process or to exact revenge for past injustices done to their people by those of other countries.

If these possibilities are taken into account, then, despite being based upon a near worst-case approach, this book may reflect a conservative, albeit pessimistic, view as to what will happen. It is also hard to fully account for the sort of ultra-extreme weather events that will begin to occur in 20–30 years' time. Nor can what happens when mega-tornados or super-hurricanes hit nuclear power plants be known. We obtained some indication of the extent of the possible release of radiation and radioactive pollutants when a tsunami hit Japan's nuclear power plant in Fukushima in 2011. But it was just an indication of what might happen. And radioactive material is still leaking from Fukushima. So that story is not over.

Nor can anything be known about what will happens when massive numbers of species we do not rely on for food become extinct. We know that flowering plants often need insects for pollination; we cannot be certain as to what happens when large numbers of insect species become extinct. Insects appear to be hardy creatures and not enough is known about the self-regulating mechanisms at work in ecosystems. We do not know what would happen, for example, if earthworms started dying off in significant numbers. Insects can be helpful, but they can also be destructive. The spread of malaria-carrying mosquitoes away from the equator as the planet warms is one example of this problem. And we just do not know exactly what happens if the birds that control the populations of harmful insects start to die off.

This book has seven chapters. The first addresses two, largely unrelated, matters. The first of these is 'the politics of science'. This refers to the process of knowledge production and reproduction in scientific communities, which results in the production of powerful consensuses as 'truth', and not the production of Absolute, or capital 'T', 'Truth'. The second matter addressed in the first chapter is 'the science of politics'. This refers to an understanding of politics as the science of the production of order, or the science of the maintenance of the status quo. Both have played and will play roles in the politics of the final hundred years of humanity.

Chapter 3 is a discussion of seven ideological factors that ensure that the trajectory toward the beginning of the final hundred years of humanity is inevitable. Chapter 4 contains an account of the climate change denial industry that includes a discussion of its members' response to the reality of catastrophic human-caused environment change. Chapter 5 is a discussion of the national level of the politics of the final hundred years of humanity. At this level, politics concerns adjusting the protection and distribution of the national population. Chapter 6 takes up the international level, looking at questions associated with migration and resource and border conflicts. The final chapter offers reflections on the politics of two of the most important of the desperate measures that will be adopted during the final hundred years of humanity: radical population reduction and stratospheric aerosol injection.

## References

Arora, N. K. (2019). Earth: 50 years challenge. *Environmental Sustainability*, 2, 1–3. https://doi.org/10.1007/s42398-019-00053-5.

Berkeley Earth, (n.d.). Global warming pause? http://berkeleyearth.org/global-warming-pause/. Accessed 2 June 2019.

Edwards, L. (2010, June 23). Humans will be extinct in 100 years says eminent scientist. *Phys.org*. https://phys.org/news/2010-06-humans-extinct-years-eminent-scientist.html. Accessed 2 June 2019.

Fischetti, M. (2017, January 4). No pause in ocean warming: scientists, not politicians, resolve a set of controversial measurements. *Scientific American*. https://www.scientificamerican.com/article/no-pause-in-ocean-warming/. Accessed 2 June 2019.

IPCC, (2018, 6 October). Global warming of 1.5°C: summary for policymakers. 48th Session of the IPCC, Incheon, Republic of Korea, http://report.ipcc.ch/sr15/pdf/sr15_spm_final.pdf. Accessed 2 June 2019.

Leahy, S. (2018, October 7) Climate change impacts worse than expected, global report warns. *National Geographic*. https://www.nationalgeographic.com/environment/2018/10/ipcc-report-climate-change-impacts-forests-emissions/. Accessed 2 June 2019.

Mizokami, M. (2019, May 20). The U.S.'s Aging 'Nuclear Coffin' Is an Ecological Disaster Waiting To Happen. *Popular Mechanics*. https://www.popularmechanics.com/military/weapons/a27533068/runit-island-nuclear-dome/. Accessed 29 July 2019).

Monbiot, G. (2007). *Heat: how we can stop the planet burning*. London: Penguin.

Pacifici, M., Visconti, P., Butchart, S., Watson, J., Cassola, F., & Rondinini, C., (2017). Species' traits influenced their response to recent climate change. *Nature Climate Change*, 7, 205–208. https://doi.org/10.1038/nclimate3223 http://www.nature.com.libproxy.murdoch.edu.au/nclimate/journal/v7/n3/full/nclimate3223.html.

Rhodes, C. J. (2019). Only 12 years left to readjust for the 1.5-degree climate change option–says international panel on climate change report: current commentary. *Science Progress*. https://doi.org/10.1177/0036850418823397.

Romm, J. (2015). *Climate change: what everyone needs to know*. Oxford University Press.

Scherer, G., (2012, December 6). Climate science predictions prove too conservative. *Scientific American*. https://www.scientificamerican.com/article/climate-science-predictions-prove-too-conservative/. Accessed 2 June 2019.

Spikins, P.A., Rutherford, H.E. & Needham, A.P. (2010). From homininity to humanity: compassion from the earliest archaics to modern humans. *Time and Mind*. 3(3), 303–325. https://doi.org/10.2752/175169610X12754030955977.

Steffen, W., Rockström, J., Richardson, K., Lenton, T. M., Folke, C., Liverman, D., … & Donges, J. F. (2018). Trajectories of the Earth System in the Anthropocene. *Proceedings of the National Academy of Sciences*, *115*(33), 8252–8259. https://doi.org/10.1073/pnas.1810141115.

Torres, P. (2016). Apocalypse soon? How emerging technologies, population growth, and global warming will fuel apocalyptic terrorism in the future. *Skeptic* (Altadena, CA), 21(2), 56–63.

Wignall, P. B. (2019). *Extinction: A Very Short Introduction*. Oxford University Press.

Youngsteadt, E., López-Uribe, M. M., & Sorenson, C. E. (2019). Ecology in the Sixth Mass Extinction: Detecting and Understanding Rare Biotic Interactions. *Annals of the Entomological Society of America*, 112(3), 119–121. https://doi.org/10.1093/aesa/saz007.

# Chapter 2
# The Politics of Science and the Science of Politics

Because the IPCC was set up explicitly to provide the major governments of the world with the scientific basis for policy action, scientists generally expected governments to follow their advice. Governments did not (Romm 2015, p. 77).

One of the first questions that arise with respect to the maintenance of a trajectory that leads to the final hundred years of humanity concerns why those in so many of the world's governments have failed to act to address the causes of catastrophic human-caused environment change. In this chapter this question is approached via a consideration of the operations of science and politics. At least part of the explanation for government inaction is that the science around climate change has been rendered sufficiently uncertain in the minds of enough people to allow those in positions of political power in many countries, especially western countries, to fail to adequately address the causes of catastrophic human-caused environment change.

Given the overwhelming number of climate scientists who affirm both the reality of global warming and that it is being caused by human activity, however, the explanation for political inaction cannot solely relate to the effects of a small number of scientists who are skeptical with respect to global warming. This leads to a consideration of the role of politics in explaining inaction with respect to addressing the causes of catastrophic human-caused environment change. Here the discussion goes to explaining why those in positions of political power are predisposed to inaction when it comes to introducing changes that would profoundly disrupt social order, which are the sorts of changes required to adequately address the causes of catastrophic human-caused environment change.

This chapter is in three parts. The first outlines a history of inaction with respect to addressing the causes of catastrophic human-caused environment change. The immediate question that arises in this context concerns why, if so many scientists are so certain that humans are causing catastrophic changes to the environment that threaten human existence (or, at least, the way humans currently exist), nothing meaningful has been done to address catastrophic human-caused environment change.



I. Cook, *The Politics of the Final Hundred Years of Humanity (2030-2130)*,
https://doi.org/10.1007/978-981-15-1259-9_2

Two intertwined explanations for this are developed in the following parts, with additional reasons for inaction being presented in the following chapter. One thread, taken up in the second part of this chapter, relates to the vulnerability of science and scientists because of the overstatement of the capacity of science to discover truths about the world. The other thread, taken up in the third part of this chapter, relates to the fundamentally conservative nature of governments, especially those in advanced industrial nations in which neoliberalism has flourished. Political inaction is, in part, a reflection of the fact that those in positions of political power seek to preserve the social order to preserve their dominant position. But it also reflects the fact that the dominance of neoliberal ideology has meant that those in government seek to preserve a capitalist market order that can neither adequately value the environmental destruction that occurs during production, distribution and consumption nor take that value into account in market transactions.

## Part One: Political Inaction

Even though climate change was raised as a threat in the 1930s and then again in the late 1960s (Boehmer-Christiansen 1994a, p. 140), the history is that of consistent political inaction with respect to addressing the causes of catastrophic human-caused environment change. Nobel Prize winner and former senior vice president and chief economist of the World Bank, Joseph Stilgitz, summed up this continuing failure to address the causes of catastrophic human-caused environment change when he wrote that despite "widespread agreement that climate change represents an existential threat… [and a] broad consensus over the urgency of action… there has been little progress … [and] the voluntary measures taken by various countries simply do not add up to what is needed" (Stiglitz 2015, p. 29). Stilgitz made those observations prior to the, allegedly more successful, Paris Agreement of 2015.

Yet, as Christoff has argued, the goals set in the Paris Accord "remain aspirational" because the agreement includes no "specific targets or dates, or the means for coordinating national contributions to ensure effective collective outcomes" (Christoff 2016, p. 781). At best, he suggested, the Paris Agreement was "a promissory note", as "its powers in ensuring outcomes are limited…" (Christoff 2016, p. 781). It is hard to dispute Christoff's conclusion that "Paris will be condemned as the conference that offered the last delusion of hope that we are tackling global warming… [and as] a missed opportunity – probably the last one – to do so without exceptional and risky technological interventions to produce 'negative emissions'" (Christoff 2016, p. 781).

The illusion of action when none has been taken has meant that, while 40 countries had implemented or were to introduce planning carbon pricing regimes in 2015, these regimes applied "to just over 10 percent of total global emissions due to exemptions and special cases" (Saha 2015, p. 50). A study published in 2018 found that 64 new laws and policies associated with mitigating climate change were introduced in 2016 and only 36 in 2017.

This might be taken to mean that the work to slow or prevent catastrophic human-caused environment change had largely been done. This inference is invalid, however, because, as the authors of the study pointed out, "the Paris Agreement requires countries … to ratchet up their ambitions over time, [which is] necessary for keeping the rise in global mean temperature well below 2°C." The Agreement not only requires additional laws and policies that revisit, revise and strengthen those that have already been introduced but also requires new provisions for "monitoring, reporting, and verification in order to comply with the Paris Agreement" (Nachmany and Setzer 2018, p. 1).

This is not to suggest that nothing is being done to address $CO_2$ emissions, but that what has been done is not enough. Indeed, the rate of growth of $CO_2$ emissions was held to close to zero between 2014 and 2016 (Le Quéré et al. 2018, p. 2160). But the rate of growth increased in 2017 and was projected to be around two and a half percent in 2018 (Le Quéré et al. 2018, p. 2413). The problem with this is that $CO_2$ emissions must *decrease* if the increase in the global average temperature is to be kept below 2 °C. And then there is the problem of removing the $CO_2$ that remains in the atmosphere and of addressing the other aspects of catastrophic human caused climate change that are threatening to bring on the final hundred years of humanity.

While there is not complete unanimity amongst climate scientists, a study published in 2013, found that over 97% of papers published on the subject in authoritative climate science journals supported the view that humans were causing global warming. As the authors of this study put it "the number of papers rejecting the consensus on [human-caused global warming] is a vanishingly small proportion of the published research" (Cook et al. 2013, p. 1). A later examination of a sample of the 2% of papers whose authors rejected human-caused global warming found a pattern of flaws in the method and mistakes in these articles (Benestad et al. 2016, p. 699).

And, though climate change skeptics dispute it, climate scientists continue to publish (what they consider to be) evidence that climate change is not being adequately addressed and its effects on the planet are increasing. Measurements of the global sea level in 2017 indicate that it has risen by "more than 7 cm… over the last 25 years" (Nerem et al. 2018, p. 2022). This, for Ablain et al. is "one of the key indicators of climate change" (Ablain et al. 2017, p. 8). The evidence that has been compiled means that the view that the average global surface temperature is increasing due to human activity is the consensus amongst an overwhelming majority of climate scientists.

Yet nothing meaningful has been done, nothing meaningful is being done and nothing meaningful will be done. So, the problem of catastrophic human-caused environment change has been established in the minds of almost all climate scientists and is accepted by the leaders of most, but not all, countries. And most people in all countries of the world accept that a significant and pressing problem exists. So, the need for urgent action has largely been acknowledged and some action has been undertaken in the name of addressing catastrophic human-caused environment change. But it is not enough to avert the onset of the final hundred years of humanity.

And at least some of the explanations may be found in the politics of science and the science of politics.

## Part Two: The Politics of Science

This part begins with a discussion of the promise of truth that is so often taken to be fundamental to science or being scientific. This promise is based on the practices of science being presented as objective, dispassionate and neutral. Each of these claims is an over-representation of both what science is and what it can be. But these over-representations have always been useful because they enhanced the status of scientists and science (and positioned science to demand the lion's share of research funding). The problem is that, at most, science produces powerful consensuses amongst scientists that work when dissenting voices are silenced and do not work when those voices cannot be silenced.

This problem has a profound effect, as will be discussed in the second part of this chapter, when those in positions of political power in modern societies are predisposed to preserve the status quo and ignore warnings from the overwhelming number of scientists who participate in the consensus that human activity is causing catastrophic changes to the environment. The politics of science and the science of politics are brought together at the end of this chapter. Here the IPCC, as an intergovernmental panel, is understood as both a scientific body and a political institution, with its politics having a very significant effect on its predictions (which explains why IPCC reports were not used in developing the scenario upon which the discussion in this book is based).

### *Science and the Promise of Truth*

When it comes to the first thread, the politics of science, the problem of inaction is a consequence of the representation of science as the source of absolute truth, objective truth, or 'capital "T" Truth'. For adopting the position that science produces Truth has made climate scientists vulnerable to challenges from a relatively small number of such dissenting scientists who either deny that the average global surface temperature has increased or deny that any recorded increase is a result of human activity. This has left those climate scientists who believe that the average global surface temperature has increased and that this is a result of human activity open to being discredited in the eyes of enough voters, politicians and policy makers to ensure that their science is largely ignored by politicians and policy makers in many, and many important, countries.

The problem begins with the very notion of Truth, first because "capital 'T' Truth denotes divinity or some essential truth..." (King 2001, p. 189). Scientists may not claim divinity, but they demeanor of many scientists can suggest that they have access to essential truth. This access does not depend on the character of the scientists who bring us the truth, however, as "capital-T Truth is an objective property of statements, something to be discovered, not created. … The Truth (with capital T) of a statement S consists in its correspondence with objective (i.e., S-independent) reality" (Mäki 1995, p. 1306).

The elision of the personality of the scientists who speak the truth means that, from this perspective at least, "science possesses a universal validity which is in principle independent of the people, society, and interests among which it happens to have sprung up, so that sciences are merely systems of formal propositions and procedures for the construction and corroboration of theories" (Suvin 2010, p. 82). As Herschel wrote in 1851, "the grand and indeed only character of truth is its capability of enduring the test of universal experience, and coming unchanged out of every possible form of fair discussion" (Herschel 1851, p. 10). Such a view of science "eliminates the knowing subject, individual and collective, in favour of an artfully posited 'objectivity'… the ideal scientist must eliminate all of his senses, except 'part of an eye that he might observe'…" (Suvin 2010, p. 82).

The power of science lies partly in its projection of a capacity to identify absolute, objective, universally valid (and, following King's lead, divine and therefore eternal) Truths and partly in its projection of the independence of this Truth (at least with respect to the scientists who deliver it). The most powerful attribute of science is the claim that it delivers the Truth, or is claimed to discern the Truth, in a manner that has nothing to do with the values or views of the scientists who announce it. This means that the force of Truth impels agreement amongst scientists and not the force of personality or a predisposition to accept a specific truth as True.

## *The Ideal Scientist: Objectivity, Neutrality and Dispassion*

The image of the 'man of science' emerged in the seventeenth century, but it wasn't until the nineteenth century that 'he' took on the characteristics commonly associated with the persona of 'the scientist'. The first characteristic of the scientist, which applied from the outset, was that he was a man. "[T]he 'man of science' was a form of masculinity: a cultural type composed of highly gendered characteristics and virtues" (White 2016, p. 158); with women's participation in science being constrained, downplayed and ignored (Schiebinger 1987).

The 'man of science' was also "a moral figure standing outside politics and above private interests, and therefore able to provide knowledge that was truly beneficial to all of society" (White 2016, p. 156). Aspects of this morality derived from religion (principally Christianity). The 'man of science' was not, at that time, an atheist. "As they gained prominence in education, government committees, and the periodical press, men of science became part of a broad community of learned elites who saw themselves as the collective bearers of natural knowledge and Christian culture" (White 2016, p. 157).

As White noted, the 'man of science' is not precisely the same person as the modern scientist. Important continuities between the two exist, however, in particular "the public image of purity, and the more private sense of a vocation whose main motive is truth." At first, maintaining these were part of an individual journey for each scientist. By the middle of the nineteenth century, however, the ethos of the scientist was taken to emerge from "a highly formalized and impersonal system of training … relocated within new research institutions that defined themselves as havens from commercial and political interests." (White, p. 161). These institutions protected "pure science" and "were typically made from institutions financed by wealthy donors or lucrative government contracts. Once they obtained the job security and institutional prestige that men of science had lacked, modern scientists became a more insular community, with closed meetings and peer review, and a highly technical language comprehensible only to the specialist" (White, p. 161).

The most powerful obstacle to finding the Truth is prejudice or bias. Or put more generally, the obstacle to discovering the Truth is having an interest in the outcome of a scientific inquiry. We believe that scientists deliver the Truth because we believe that they are telling us The Truth and not their truth (and what they want us to believe). People have faith in science because we believe that scientists do not offer explanations (Truths) because they serve the interests of the scientists who provide them, but because they are the Truth. More importantly we believe that they tell us the Truth because that is what they have been trained to do; that disinterestedness is central to the attitude a scientist must develop to be a scientist.

The important thing is that scientists see themselves as remaining neutral about what they observe and take into account as evidence when seeking to explain the phenomena that they study. They are to set aside every motivation other than the motivation to find the Truth. Above all, they are not concerned with the consequences of their discoveries for society or politics. "The scientist does not, as a matter of vocation, campaign for social and political reform, or engage in moral or religious debate, unless perhaps to defend the very purity of science itself" (White, p. 161).

Once again, an important aspect of the image that scientists want to project is that they are always neutral and always seek to avoid bias, and for a very long time that is what most people thought scientists did. The scientist as neutral observer was the image they projected of themselves and we accepted this. The only passion that scientists admit to is a passion for truth. Otherwise, they claim to be dispassionate and present their truth as something that just comes from their research. This way of presenting themselves has been very effective for scientists in legitimizing their research and in persuading others, especially those in government, to fund it.

But objectivity, neutrality and a dispassionate attitude are difficult to achieve and impossible to maintain. It is useful to treat objectivity, neutrality and dispassion as goals. It is important to make them into ideals to which scientists (indeed, any thinker) should aspire. Maintaining an open mind and endeavoring to avoid as much prejudice as we can are always desirable. Even if it is an impossible goal, as will be argued below, seeking to be objective, neutral and dispassionate remains a laudable goal.

## *Scientific Truth and Consensus*

But this way of representing themselves has never been an accurate reflection of what scientists do. This is not the place to present an extended discussion of philosophy and philosophy of science. The important point here is that objectivity, neutrality and dispassion are not possible, even when they are desirable. It is desirable that we commit to seeking to achieve these states of mind. The problems come when we begin to believe that we are objective, neutral and dispassionate, when we convince others that this is how we are and, therefore, that what we present as the Truth is free from our subjectivity, bias and passion.

For, in the end, science is a social activity and, if the term wasn't used exclusively as a pejorative, could be thought of as a form of 'groupthink'. Scientists adopt ways of seeing and studying phenomena that reflect their membership of a group. Others had identified this feature of science before him, but one of the best-known scholars to develop this understanding of science was Thomas Kuhn, who referred to this as scientists participating in a paradigm (Kuhn 2012). Paradigms, for Kuhn, determine both the way that observations are made and the making sense of the observations in which members of different scientific disciplines share. In *The Structure of Scientific Revolutions* Kuhn "used the then-obscure word paradigm to refer to the shared ideas and concepts that guide the members of a given scientific field" (Orman 2016, p. 47). To belong to the group, that Kuhn referred to as "normal science" is to share in the paradigm, and the group rejects those who reject the paradigm.

> Following Kuhn, when a field is in a normal science state, the fundamentals are taken for granted, and those scientists working in the field are able to concentrate on very specific problems. And scientists working in such a field tend to focus on problems that appear to be easily solved with the conceptual resources supplied by the reigning paradigm or theory. Indeed, these conditions are what make possible a rapid succession of discoveries, the sort of pattern we associate with the accumulation of knowledge (Wray 2018, pp. 647–8).

Problems arise, however, when observations begin to accumulate that cannot be reconciled within the paradigm being applied in and as normal science. Those working within the paradigm may tend to ignore or downplay the significance of such observations until other, 'revolutionary', scientists offer a new paradigm. The crucial question, of course, concerns the conditions under which revolutions in science occur. And the answer is not a simple one.

Akerlof and Michaillat used case studies from astronomy, medicine, and economics to develop a model in which the 'power' of inferior paradigms meant that they were maintained despite the existence of superior paradigms. "In these case studies, power is determined both by the data and methods available to test theories and by the norms regarding the appropriate criteria to use in tenure decisions" (Akerlof and Michaillat 2018, pp. 13228–9).

Akerlof and Michaillat's model showed that "inferior paradigms prevail when they are in contest with better ones... [when] scientific tests lack power, or are little used in determining admittance into the fellowship of established scientists…". In these cases, "the chances of getting trapped in an inferior paradigm are high." Even though their model focused on the granting of tenure "the model could describe other aspects of the scientific process, such as hiring of junior faculty, award of grants and honors, and publication in scientific journals [and] … hiring committees, award committees, and journals' peer-review systems" (Akerlof and Michaillat 2018, pp. 13232–3).

Paradigms are everywhere and not just in science, and they are probably inescapable. Members of religious groups share in paradigms. But it is not just believers of any kind for whom paradigms are important. Members of a society can share a paradigm. The fact that few people can imagine societies functioning very differently from how they function now is one of the main factors behind the maintenance of a path that leads to the final hundred years of humanity. It is, as many have noted, easier to imagine the end of the world than it is to imagine the end of our way of life. Our ways of life make sense to us. And, from where we stand, doing things differently makes no sense. But what makes no sense to us provides clues for identifying the paradigm within which we operate.

Scientists, like everyone else, work within a paradigm and they usually confirm the paradigm that they share with other scientists. Almost all scientists do what Kuhn called 'normal science.' But because they remain committed to objectivity, neutrality and dispassion even those doing normal science remain open to having their paradigm disrupted by other scientists whose observations do not fit the paradigm or who develop theories that use a different paradigm to make sense of what scientists have discovered through observation and/or experiment.

The best recent examples of the way that scientists are constrained by a paradigm are to be found in the emergence of relativity and quantum mechanics. One of the important theorists of relativity, Albert Einstein, was thinking about time and space in the usual (Newtonian) way when he started to think about gravity as a 'bending' (or changing) of space and not as a force that one body exerted on another body. This ultimately led him to develop his theory of general relativity. The problem Einstein and other physicists of his time were addressing was that observations were undermining the Newtonian paradigm, which had successfully guided physicists' mode of interrogating the world for centuries. For those working within the Newtonian paradigm had "predicted every gravitational observation for over 200 years" (Robinson 2018, p. 1). General relativity theory resolved the problems with this paradigm. But only by subverting it.

The Newtonian paradigm had always had a problem in being incapable of explaining the physical process that gave rise to gravity (see Kochiras 2009). This wasn't much of a problem, in the sense of requiring that the Newtonian way of thinking be revised. But 200 years later "it was shown to have an error of $\approx 1.2 \times 10^{-8}$ in its prediction of Mercury's orbit" (Robinson 2018, p. 1). This provoked more systematic attempts to resolve the problem with the Newtonian paradigm. The problem was that it required a new theory: Einstein's general relativity theory.

The result was a revision of the fundamental mode of thinking that had been supporting physicist's work for centuries. By way of replacing the Newtonian paradigm, "Einstein showed that mass distorted the fabric of space–time, producing gravity. He gave a mathematical description of his calculations through his gravitational field equations." Most importantly, for this discussion however, Einstein "solved the difference in Mercury's orbit and … went on to make other predictions, some of which have been verified" (Robinson 2018, p. 1).

The development of quantum mechanics did not affect many scientists other than those engaged in physics, as most scientists work with objects larger than atoms. For those scientists working with sub-atomic particles, however, nothing was the same. The behavior of sub-atomic particles was so unlike the behavior of larger particles that a strange new world was discovered to exist at a level below the one that people experienced. One of the more disturbing aspects of quantum mechanics is the level of the randomness of the behavior of sub-atomic particles.

Even though physicists could predict the behavior of larger particles and objects, they discovered that they could not always predict the behavior of sub-atomic particles. And being wrong some of the time meant that they did not have the Truth about those particles. Einstein himself found this objectionable and refused to accept the unpredictability, or uncertainty, that lay at the very core of the physical world that quantum mechanical theory implied (and sometimes quantum mechanics discovered predictability that did not make sense – see the 'double slit experiments').

That there could be something like laws of physics that had uncertainty at their core was anathema to Einstein. He believed "that natural laws could not be like the throw of dice, with the inherent randomness or probability. But this is exactly what Quantum Mechanics tells us – that at the fundamental level Nature is inherently random, codified in Heisenberg's famous Uncertainty Principle." (Natarajan 2008, p. 656). Heisenberg's "famous Uncertainly Principle" holds that it is impossible to know in the same moment both where a sub-atomic particle is and how fast it is moving. This means that some measurements cannot be made certain in quantum mechanics, and the behavior of sub-atomic particles can only be described in terms of probabilities.

Quantum mechanics met with resistance when it first emerged and challenged the paradigm within which physicists had worked and were working. As noted, Einstein himself struggled with an approach that seemed to him to undermine the very idea of a scientific law. Up to this point Einstein and almost all the other physicists had been engaged in normal science and had been working with and within a Newtonian paradigm. This is what is intended in the reference to the politics of science in the title of this chapter. Scientists who think outside the paradigm challenge a group's way of thinking about and doing science. Their ideas will be considered, if they are considered at all, to be poor science by most of the other scientists in their field. Usually the other scientists will be right and the maverick wrong. Occasionally the mavericks will be right.

The real strength of science is that all scientists' commitments to objectivity, neutrality and dispassion mean that they are always open to being persuaded to think differently. They might fight hard to hold onto the paradigm within which they have always worked, but they will not always win. If the evidence mounts and the theories can be developed to explain that evidence (and these theories are the new paradigm) then, no matter how much they want to hold onto their old ideas, scientists must give in to the new way of thinking or stop being scientists in that field.

The problem arises not so much with how scientists conduct themselves than it is with how they represent themselves or allow themselves to be understood. In short, the problem lies in how we conceive science. As Oreskes and Conway put it: "We think that science produces certainty, so if we lack certainty, we think the science must be faulty or incomplete" (2011, p. 267). Yet, they add, "history shows us clearly that science does not provide certainty. It does not provide proof. It only provides the consensus of experts, based on the organized accumulation and scrutiny of evidence" (Oreskes and Conway 2011, p. 267–8).

The point of this discussion is not to discredit science or to imply that everything is relative and one person's opinion as to how the world is cannot be better than someone else's. While great care is needed in this context, we must recognize that there is no outside of a paradigm, so there is no position from which to judge one paradigm that is not already part of another paradigm. Acknowledgement also needs to be made that the best that can be done is to be as careful as possible when it comes to deciding what is true. No one has the Truth but some people have positions that are much more likely to be true, or much closer to the Truth (to which we can never get) than the positions held by others.

People who are committed to being objective, staying neutral and remaining dispassionate are much more likely to be closer to the Truth than people who refuse to subscribe to these (impossible) ideals. So, even though we know that science is a group activity, we must trust the people in this group more than we trust those in most other groups. All but a tiny fraction of scientists work diligently and conscientiously to conduct the best science they can. We can be aware that their membership of a group limits what they can imagine and study, and that they even reject theories that might turn out to be true. But this does not mean that we have grounds for not trusting that they will usually be right and that they will eventually get it right if they are provided with evidence that undermines their paradigm and they develop theories that offer a new paradigm that makes sense of the observations that they make.

To reject the scientific consensus that catastrophic human-caused environment change is occurring and threatens to bring about the end of humanity is a poor option. As Oreskes and Conway argued, "sensible decision making involves acting on the information we have, even while accepting that it may well be imperfect and our decisions may need to be revisited and revised in light of new information." And we must acknowledge that, "even if modern science does not give us certainty, it does have a robust track record" (Oreskes and Conway 2011, p. 273). So not trusting the scientific consensus with respect to catastrophic human-caused environment change is not defensible. It is not a case of blind trust. "But without some degree of trust in our designated experts – the men and women who have dedicated their lives to sorting out tough questions about the natural world we live in – we are paralyzed, in effect not knowing whether to make ready for the morning commute or not" (Oreskes and Conway 2011, p. 273–4).

Before leaving this discussion, it is important to note that the level of agreement as to the earth's climate changing because of human-caused environment change amongst earth scientists is very high. It is not quite the 97% that is often quoted, this number relates to the percentage of articles published in scientific journals (Wiens

2016, p. 71). Journal articles are accepted after members of the group (who share in the paradigm) have reviewed them. So, a selection bias is at work here.

As Bauer argued, "peer review… is not necessarily more objective than is literary criticism or art criticism, it is just the expressing of opinions by interested people with supposedly appropriate expertise. That expertise is judged by and naturally reflects standard contemporary beliefs, and thereby it tends to entrench whatever happens to be the contemporary consensus" (2017, p. 4). But peer review is central to the way that all sciences function, so most scientists are committed to validity of the peer review process. And even if "peer review itself is a social construction, not a 'truth machine', … [it is] one more likely to produce a closer approximation to 'reality' than any other epistemology" (Schneider 2001, p. 341).

A survey of a sample of members of the American Association for the Advancement of Science conducted by the Pew Research Centre in 2015 found that 87% of members of the Association believed that climate change was mostly human-caused (with 93% of scientists who were working and who held degree in Earth Science agreeing that human activity was the main cause of climate change) (Pew Research Centre 2015).

The problem that scientists created for climate scientists was and is that they presented themselves as having the right disposition (as men of science who became modern scientists) to provide the Truth in their field. The Truth, as has been discussed, is true for everyone and anyone who can claim to be an expert in a specific science should, as a result, accept this Truth. The notion of a 'maverick scientist', who is qualified but denies the Truth accepted by almost all the other scientists in their field, makes little sense when you have this view about science. And the climate change denial industry relies on maverick climate scientists who have some qualifications but deny what most other climate scientists hold to be true. The problem lies not in science, but in the way that scientists represent their endeavor. For "no bit of science is guaranteed true for all time. And it isn't 'objective' either; it is (relatively!) impersonal because so many hands have done something with or to it, but it remains a human consensus" (Bauer 2017, pp. 103–104).

The situation with respect to political inaction with respect to addressing catastrophic human-caused environment change may have been somewhat different if scientists had not created the image of the scientist uncovering the Truth through disinterested research. The next part of this chapter might be taken to suggest that political inaction would still have occurred. And it may well be that, because of who they were and what was asked of them, scientists could not have admitted that they did not have the Truth but had the very best guesses produced by the very best thinkers whose guesses are continually subjected to testing by people who remain open to the idea that what they hold to be true right now may not be the Truth. But they did not, perhaps could not, do so. And this has left climate scientists vulnerable to the views of a small number of deniers who have qualifications that mean that they might be thought to know something about climate science.

Petto, however, offered a different explanation that deserves acknowledgment here. In his view, scientists have made too much, and not too little, of the indeterminacy of science. His position, as Nowakowski and Teich explain it, is that "people question science because it is revisable and adaptable…" (Nowakowski and Teich 2015). Or, quoting Petto himself: "Part of the problem is really a problem of authority. We scientists have made a big deal about the fact that science is revisable, science is tentative, science can be changed when new information comes along…" (as quoted in Nowakowski and Teich 2015). I do not agree with Petto.

The World Wide Web and social media have not helped, in that they have enabled non-specialists to amplify the views of mavericks who would otherwise be silenced by group dynamics amongst climate scientists. But politics has been the main cause of the skepticism that climate scientists have encountered when attempting to give us their very best guesses. They may be skilled when it comes to science, but scientists are less adept, perhaps deliberately so, when it comes to politics. It also does not help that they are often poor communicators, as this also gives those who 'do' politics the upper hand.

The problems for scientists arise when a political motivation exists to discredit their work. This does not arise with respect to most forms of science, which continue to be trusted by members of the community. A 2016 study of adults in the United States by members of the Pew Research Center found that "public confidence in the scientific community appears to be relatively strong…" (Pew Research Centre 2017). But public confidence is not as high as it once was, as "only four in 10 people reported a great deal of confidence in the scientific community."

More importantly for this book, trust in some scientists, including climate scientists, is not as great as with other scientists. According to a report from the Pew Centre, "many people hold skeptical views of climate scientists and GM food scientists…" (Pew 2017). Crucial, in this context, is that "people's views about climate scientists vary strongly depending on their political orientation, consistent with more than a decade of partisan division over this issue" (Pew 2017). And this brings the discussion to politics and its relationship with science.

## Part Three: The Science of Politics

Science and politics have a complex relationship. Science enabled those in positions of political power to free themselves from the need for support from religious leaders or from those who would challenge their authority on religious or spiritual grounds. According to Bertrand Russell, during Charles II's reign (1660–85) "scientific rejection of traditional superstition became common amongst learned men. Charles II perceived that science could be an ally against the 'fanatics', as those who regretted [the demise of radical Puritan Oliver] Cromwell were called. He founded the Royal Society, and made science fashionable" (Russell 1967, p. 11).

Religion continued to be important to political rule, however. But the rise of science lead to a change in the relationship between religion and political rule, which was rendered more available because the newer religions in Europe were Protestant, which allowed a greater role for an individual's reasoning capacity. Monarchs still claimed divine justification for their rule. But "the ways in which monarchs claimed to act as mediators between the heavenly and earthly worlds underwent a crucial change. Sacred kingship was supplanted by rational rulership, within the context of a religion which laid much greater stress on individual moral responsibility".

## A New Mode of Governance

This, as Briggs suggested, resulted in new "theories of sovereignty, while the state itself accumulated both new powers and new problems." These new theories of sovereignty, and those powers and problems led to a reimagining of the role of government and the eventual emergence of a form of government (the neoliberal state) that is incapable of addressing the causes of catastrophic human-caused environment change. As noted above, in the first place the change was in the justification of monarchical power, with "sacred kingship" giving way to "rational rulership". This means that justification for monarchical power are now based on calculations as to its efficacy and not, as will be discussed below, its goodness. Adopting Foucault's terminology, in this transition political rule comes to be understood as a science of the governance of populations.

This introduces a fundamental change in the mode (paradigm) of governance. Curtis has presented this in terms of a distinction between the governance of 'populousness' and the governance of 'population'. The governance of 'populousness' involves the direction an integrated, organic community of those who fulfill specific social roles. The governance of 'population' involves the management of a collection of individuals.

As Curtis explained, population differs from populousness in a variety of crucial ways. First, "population" depends "on the establishment of practical equivalences among subjects, objects or events. In contrast to populousness, whose logic centres on the hierarchical differentiation of essences (knights fight, priests pray, peasants till), population depends upon the notion of a common abstract essence" (Curtis 2002, p. 508). Second, "population consists of so many undifferentiated atoms distributed through abstract space and time" (Curtis 2002, p. 508). Third, "population is primarily a statistical artefact. The establishment of practical equivalences means that population is connected to the law of large numbers, which causes individual variation to disappear in favour of regularity." (Curtis 2002, p. 509). Finally, though only "in its developed forms, population is bound up with the calculus of probabilities. Population makes it possible to identify regularities, to discover 'things which hold together'…" (Curtis 2002, p. 509).

An important change that has occurred through this transition from the government of populousness to the governance of population in this context is a movement away from the direction of discrete and essentially differentiated subjects (or better subjectivities) to the administration of generic or generalized individuals all of whom obey laws of large numbers. At the risk of being accused of playing word games, the difference concerns 'obedience' when it comes to the government of populousness and 'obedience' in the context of the governance of populations. In the former, 'obedience' refers to the direct manipulation of specific figures. Knight, priest and peasant are called to conform to, or obey the requirements of their position and the dictates of the monarch. In the latter, 'obedience' refers to the following of (obeying) general, if not universal, patterns. Individual (atoms) obey the law of large numbers and their regularities are not the result of conscious intention to conform to directions from those in superior positions.

A consequence of this shift from the government of populousness to the governance of populations was a loss of a reference to the morally 'good' went it came to government. Whereas the government of populousness meant the organization of discrete bodies with specific sets of expectations or requirements for their goodness, the governance of populations involves no pursuit of goodness. It makes sense to think of the good knight or good priest and even peasants could be differentiated based on their approximation of an ideal. It makes no sense to think of individuals pursuing their individuality in a good or bad way. Individuals can express their individuality more or less fully and individuality may have positive or negative consequences, though this may depend on the perspective from which those consequences are viewed. But there is no good individuality. And there is either conformity to the regularities that populations manifest or there isn't; and while conformity might be taken to be good, within this conception of government it is not a moral good.

Indeed, the very notion of a moral good has been left behind. And the principal question that arises from the loss of goodness as an objective of governing concerns what replaces goodness as its objective. What can be taken to be the modern answer to this question is: order. And this brings us to the writings of Niccolò Machiavelli, who was, for some, one of the first writers to express a modern approach to politics (Mansfield 1981) and even the first political scientist (Farr 1988, p. 52). Machiavelli is important here for two reasons. The first is that his legacy with respect to a modern approach to politics is taken to be an amoral approach to governing. This grossly oversimplifies Machiavelli's position, but it captures something important in the change in approach to governing that is part of modern governance. Indeed, in a chapter of *The Prince* Machiavelli "calls for learning how to be not good", announcing that by doing so "he departs.... 'from the orders of others'" (Mansfield 1981, p. 294). These "others" were those, such as the Christians, who "based their political science on what should be done rather than on what is done, who elaborated the 'profession of good' in all regards..." (Mansfield 1981, p. 295).

## The Science of Order (Ordering)

So if governing is no longer with reference to the good, then the crucial question concerns the purpose of those who govern. And the answer to this question was: to create and maintain order. Whereas political leaders in previous eras presented themselves as responsible for improving their subjects, modern political institutions simply manage their citizens to ensure social order. In the Middle Ages, according to Foucault, "the sovereign was commonly defined as someone who must help his subjects gain their salvation in the next world." But this ended and, ever since, government has "no longer has to concern itself with the salvation of its subjects in the hereafter…". So, "whereas in the Middle Ages the sovereign's paternal role was always very emphatic and marked", now "the state is a specific and discontinuous reality… [that] exists only for itself and in relation to itself..." (Foucault 2008, p. 4–5).

This is a crucial shift and ends with political institutions that have no role other than to preserve the status quo. To preserve the status quo is to preserve a specific social order. To preserve that social order is to preserve a specific political order or system of governance. The preservation of the system of governance, then, becomes the end of a system of governance. But the emergence of a science of politics wasn't just about being realistic, though this was central to a modern approach to governing. Mansfield sees Machiavelli as the first to "vividly and profoundly [express] … the realism and dynamism of modernity" (Mansfield 1981, p. 295). It is about being scientific. While the abandoning of the (moral) good and realism are parts of the move to a (modern) scientific mode of governance, so too is the move to approach and study government in a scientific manner.

This brings us to the emergence of those who sought both to practice and study a science of politics. In Farr's view, "the expressly articulated concept of a 'science of politics' emerges out of the… Scottish Enlightenment in the eighteenth century" (Farr 1988, 52). As has been observed, this new 'science' was for both those who undertook political practice and for those who studied politics. When it came to studying politics, the science of politics was based upon "*a Newtonian method for the study of politics*" (Farr 1988, p. 52 emphasis in original). When it came to political practice, the science of politics was "*a politics of moderation* whose tasks included a critique of 'enthusiasm' and other excesses of mind and body politic. In this way, the science of politics, the politics of moderation, and the critique of enthusiasm were conceptually connected" (Farr 1988, p. 52 emphasis in original).

The science of politics is, as we should expect from its origins in Machiavelli's ideas, fundamentally conservative. Those in political institutions preserve the status quo, which requires the preservation of the political institutions through which they govern. The "politics of moderation" that is central to the science of politics requires the suppression of any attempt at radical change, especially radical change to achieve a better society. In its later iterations, for those committed to the science of politics "ideology literally replaces *enthusiasm* as the name of the poison to be remedied by the scientific moderation of political science" (Farr 1988, p. 52 emphasis in original). Certainly, selfishness provides an explanation for the commitment that those in positions of political power have to maintaining the status quo and their position in it. But modern political institutions are also infused with a modern and scientific approach to governance in which radical change, that is change at the root, is resisted. (Until radical change is necessary to preserve the status quo, which will be taken up in Chap. 7.)

There is more to it than this, however. For the other aspect of the science of politics, Newtonian method, provides another conservative effect on governance by ensuring that it is studied, at least by mainstream political scientists, as a study of the actual. Political science came to be conceived as one of a number of "probabilistic forms of empirical and causal inquiry into social or human matters that took methodological inspiration from the natural sciences, especially Newtonian mechanics. Here at the veritable coining of *political science* we recognize a glimmer of the modern conception of the nature and limits of what have come to be called the social, human, or behavioral sciences…" (Farr 1988, p. 57). And the most important of these limits are that the study of politics is the study of what is, and not what can be or what ought to be.

The effects go further than this, however, in that Newtonian mechanics is based upon a conception of reality as constituted by discrete objects (in the way that the planets of the Newtonian conception of space-time were discrete – and not specific effects within a whole comprised of vibrating strings, which is a quantum mechanical view). This recalls Curtis's discussion of the new mode of governance based on population (and not populousness). Population, which "consists of so many undifferentiated atoms distributed through abstract space and time" and is constituted as "primarily a statistical artefact", manifests "regularities" that allow political scientists "to discover 'things which hold together'."

## *The Science of Politics and the Market*

And the principal means through which the first political scientists saw society holding together was via the market. For one of the leading thinkers of the Scottish enlightenment, in which political science emerged, was Adam Smith, the best-known advocate of the value of selfishness and of the market as a principal ordering device. Both ideas belong to a liberal tradition, of which Hobbes is part, but Smith departed from Hobbes's position that the undifferentiated atoms called human beings had to be held together, or ordered, through a powerful central authority or Leviathan (Hobbes 1946). For Smith it was the 'invisible hand' of market forces that helped to ensure that individual selfishness became both productive and progressive.

To Smith, though, the market's invisible hand needed God's intervention to ensure that morally good outcomes resulted from market processes. For the (liberal) political scientists who followed him, and for whom the supernatural was no part of science, however, selfish competition becomes the driver of economic and social order and economic and social progress. Removing "the critical coordinating role of Providence from Smith's system" later political scientists posit "the self-correcting abilities of a market economy." The result is an acceptance of the, now dominant, "neo-classical paradigm, buttressed by the denatured remnants of Smith, [that] argues that a competitive system is naturally stable unless destabilized by external intervention. Moreover, it will automatically correct disruptions and return quickly to an optimal growth pattern" (Martin 1990, p. 273).

In the new, modern (liberal-capitalist) political science, then, government is legitimate, legal and rational if it intervenes in society only "insofar as interest, or interests, the interplay of interests, make a specific individual, thing, good, wealth, or process of interest for individuals, or for the set of individuals, or for the interest of a given individual faced with the interest of all, etcetera. Government is only interested in [individual] interests" (Foucault 2008, p. 45). This is a change in the very idea of government, which is no longer about improving people and society and concerns managing their interests in the most efficient way possible, with efficiency being determined by market processes. The "fundamental question" for modern (liberal-capitalist) governments became, as Foucault suggested: "What is the utility value of government and all actions of government in a society where exchange determines the true value of things?" (Foucault 2008, p. 46).

And, once again, it is important to note that the question of the true value of things is not a moral question and cannot be answered through reference to a moral good. The new political science, which is based on the study of the behavior of "undifferentiated atoms distributed through abstract space and time" that manifests "regularities" has no place for the moral good. So, even if they were to be asked of the moral good, political scientists can offer nothing. For "on the question of whether … [political] science can arrive at judgments of what is 'good' and what is 'bad,' the answer is that it cannot—that such judgments are, indeed, the task of ethics as a separate enterprise" (Eulau 1969, p. 12).

To the extent to which there is a 'good' of modern political science, both as political practice and mode of study, it is defined by the maintenance of an order that serves individual interests. More particularly, it is the maintenance of a (liberal-capitalist) market order. In addition, political science rejects any ideology, or enthusiasm, that threatens this order. At the very moment when the trajectory of human economic development has led us to the brink of the final hundred years of humanity, the science of politics requires the preservation of the very order that is causing catastrophic environment change and the refusal of any 'grand plans' that might result in addressing the human causes of catastrophic environment change. And those who study politics are unable to criticize them for doing so.

## The Politics of Science, the Science of Politics and the IPCC

One, possibly the best, illustration of the ways that the science of politics and the politics of science contribute to government inaction with respect to addressing the human causes of catastrophic environment change are the reports and responses to the reports of the Intergovernmental Panel on Climate Change (IPCC). The members of this body have one of the most difficult roles to play in the history of science communication. First, they must negotiate a difficult terrain around their authority as scientists and, more importantly, as scientists who speak on behalf of thousands of other scientists. And, second, they must produce reports for the governments whose members support the panel. The members of the IPCC, then, must negotiate a difficult, partly political and partly scientific, terrain. And, as Hulme & Mahony suggested, "Maintaining scientific integrity and quality control, while retaining political credibility and salience – the classic twin goals of a science-policy boundary organization – is not easy" (Hulme & Mahony 2010, pp. 709–10).

The IPCC was created in 1988 by the World Meteorological Organization (WMO) and the United Nations Environment Programme (UNEP). Its task is "to provide governments at all levels with scientific information that they can use to develop climate policies." It is "an organization of governments that are members of the United Nations or WMO" and currently has 195 members. The IPCC provides Assessment Reports, Synthesis Reports and Special Reports. The first are comprehensive accounts "on climate change, its causes, potential impacts and response options" produced by three working groups, which focus on the physical science of climate change (Working Group I), adaptation and impacts (WGII), and mitigation options (WGIII). Synthesis Reports summarise the Assessment Reports. Whereas Special Reports "are an assessment on a specific issue" (IPCC).

IPCC reports are supposed to capture "scientific information" and to do so in such a way that governments can use that information "to develop climate policies." The questions that arise, then, concern the processes through which the IPCC speaks for those in the scientific community and to those in governments. It must play two forms of politics: that which is part of the politics of science and that which is part of the science of politics. Participating in both forms of politics at, more or less, the same time would prove difficult under any circumstances and particularly so in the context of monitoring and addressing climate change.

For, in both cases it is a question of the IPCC establishing and maintaining its authority, which is one of the most important of all political resources. As Porter, Kuhn and Nerlich suggested, authority is "an attribution people make about a specific actor based on either role or ability to provide reliable information; it is a source of power that allows an actor to give orders and enforce obedience" (Porter et al. 2018, p. 876). Crucially, authority is "less the possession of an individual and more a characteristic of the relationship between actors occurring in a given context" (Porter et al. 2018, p. 876). Further, as they wrote, "authority can refer to either a capability associated with a particular position in the agreed-upon organizational structure (e.g. an officer in a bureaucracy) or a person's acknowledged expertise with respect to the task at hand…"; though they added that "both resources should be viewed as relational in orientation" (Porter et al. 2018, p. 876).

The fact that these two sources of authority, expertise and position, are relational means that "both positional and expertise-based authority are, potentially, the site of negotiation: Actors can allow other actors to play (or prevent them from playing) particular roles, and actors inhabiting roles require support from others in carrying out their duties. The organization in which authority operates therefore becomes an important resource for guiding performances" (Porter et al. 2018, p. 876). The IPCC continually seeks to participate in two modes of the production of authority in two different sets of relationships. In a sense, it is two organizations in one.

Oreskes and Conway offer an interesting account of the drafting of Chapter 8 of 1995 IPCC Working Group I Report (AR2 WGI). The chapter dealt with global warming. Its convening Lead Author, Benjamin D. Santer, presented the findings in the chapter to participants in plenary meeting of Working Group I. "The chapter was immediately opposed by the Saudi Arabian and Kuwati delegates. In the words of the New York Times's, these oil-rich states 'made common cause with American industry lobbyists to try to weaken the conclusions emerging from Chapter 8'" (Oreskes and Conway 2011, p. 204).

The findings as Santer presented them were rejected by these delegates even though "the IPCC … bent over backwards not to use alarmist terms." Indeed, the first chairman of the Intergovernmental Panel IPCC (1988 to 1997), Bert Bolin had "deliberately imposed a policy of extreme conservatism of language…" (Oreskes and Conway 2011, p. 206). Yet even this was not enough. "The opposition of the Saudi and Kuwati delegates… ensured only least common denominator statements. Everyone involved … [saw] how the process led to a conservative estimation of the threat" (Oreskes and Conway 2011, p. 206–7).

## *Politics of Science and the IPCC*

Before moving to discuss the politics of science and the IPCC, the power of the scientists of the IPCC must be recognised. In a situation in which paradigms determine scientific practices and the research agenda, the centrality of the IPCC in the development of a paradigm through which climate change is understood must be acknowledged. If nothing else this is because it leads to the recognition that the members of the IPCC seek to communicate a truth through their reports. In short, the intention behind their reports is to promote the consensus that human-caused climate change is happening and is happening at a rate at which it threatens human societies. The problem is that promoting a consensus is a political act rather than a scientific act and great care must be taken not to exclude views that do not conform to that consensus or to over-represent the force of a consensus. The questions of which scientific disciplines are included when teams of researchers, such as the Working Groups, are formed by the IPCC and from where these scientists come are equally important political matters that arise with respect to the IPCC.

The IPCC has become the preeminent, if not sole, dispenser of scientific information about climate change. According to Hulme and Mahony, "the practices of climate change knowledge production developed by the IPCC have altered scientific practice, not only in the biogeophysical sciences and social sciences, but also in the design of interdisciplinary work around climate change" (2010, p. 713). Therefore, they support vigilance with respect to "the ways in which international knowledge institutions like the IPCC gain power and influence in international deliberations and yet are not always open, democratic or accountable in their own modes of operation" (Hulme & Mahony 2010, p. 710). Problems with the mode of operation of the IPCC, they suggested, were revealed during controversies concerning alleged errors in the IPCC's Fourth Assessment Report. They took this incident to render prescient Shackley's warning (in 1997) of the "danger(s) of the IPCC peer reviewing process becoming too self-contained and insulated from criticism at the paradigm level" (Hulme and Mahony 2010, p. 710).

Maintaining a consensus, and the paradigm from which it results, requires ensuring a consistent position is articulated. This has certainly been the case with the IPCC. For Porter, Kuhn and Nerlich, "IPCC reports have maintained one very important claim: a scientific consensus (95% confidence) exists that more than half of the observed temperature increase from 1951 to 2012 was caused by anthropogenic (human) activity" (2018, p. 875). Hulme and Mahony observed that "since its origins, the IPCC has been open and explicit about seeking to generate a 'scientific consensus' around climate change and especially about the role of humans in climate change" (Hulme and Mahony 2010, pp. 710–1). Boehmer-Christiansen goes somewhat further by presenting "the emergence of the global warming threat as created by the international research community in its attempt to influence policy by tendering 'scientific' advice" (Boehmer-Christiansen 1994b, p. 185).

The questions that arise in this context concern how this consensus is created and maintained. Hulme and Mahony rejected an understanding of consensus creation in the IPCC as "a process of 'truth creation'… which marginalizes dissenting voices – as has frequently been portrayed by some of the IPCC's critics…". But they acknowledged that "seeking consensus can be as much about building a community identity… as it is about seeking the 'truth'" and "as Yearley explains, IPCC consensus-making is an exercise in collective judgement about subjective… likelihoods in areas of uncertain knowledge" (Hulme and Mahony 2010, p. 711). In this, the political project of the scientists involved becomes important to their science. For, "consensus-making in the IPCC has been largely driven by the desire to communicate climate science coherently to a wide spectrum of policy users – 'to construct knowledge'." The problem is that "in so doing communicating uncertainties has been downplayed" (Hulme and Mahony 2010, p. 711).

Another problem that arises in the processes of consensus making within the IPCC is that it is, by default, conservative. Projections with respect to temperature increase in IPCC's reports have been regularly criticized as underestimations. With the author of an article in *Scientific American* suggesting, in 2012, that "a growing number of studies on the topic" had found that "across two decades and thousands of pages of reports, the world's most authoritative voice on climate science has consistently understated the rate and intensity of climate change and the danger those impacts..." (Scherer 2012).

The same accusation of consistent conservatism in IPCC reports has also been directed at its projections with respect to sea-level rise. "Was the IPCC AR4 too conservative in reaching its consensus about future sea-level rise? Many glaciologists and oceanographers think they were, leading to what Hansen attacks as 'scientific reticence'" (See Hulme and Mahony 2010, p. 711). Others defended the IPCC's conservative approach in AR4, which it has continued in subsequent reports, on the ground that this is a complex issue and presenting less-conservative predictions would be improper. They argued that, instead of "reaching a premature consensus, the AR4 report stated that in fact no consensus could be reached on the magnitude of the possible fast ice-sheet melt processes that some fear could lead to 1–2 m of sea-level rise this century. Hence these processes were not included in the quantitative estimates" (Hulme and Mahony 2010, p. 711).

Another question that arises with respect to a consensus concerns those who participate in the formation of that consensus. The Working Groups created by the IPCC may be relatively cohesive, but even here some voices might be rendered weaker or excluded. So, exclusions can occur within Working Groups. "Caseldine et al. for example, concentrated specifically on how palaeoclimate research has been represented in the two decades of IPCC reports, welcoming the greater prominence given to such research in the IPCC Fourth Assessment (AR4)" (Hulme and Mahony 2010, p. 707). The problem can increase when it comes to interactions between Working Groups. Stocker and Plattner, who had served on IPCC Working Groups, suggested that "the interaction between IPCC working groups has long been challenging. Different communities have differing philosophies, approaches and terminologies, and mismatched time constraints..." (Stocker and Plattner 2014, p. 165). The politics of science may well be inescapable, and the IPCC does not escape it.

Nor does the IPCC escape the politics associated with which scientists are invited to participate in Working Groups and which are excluded. For concerns have been raised with respect to the over-representation of Western and Northern scientists. Boehmer-Christiansen, for example, accused "a 'Northern' science bureaucracy… [of] extracting data and knowledge from a globally dispersed research base increasingly dependent on multinational research funding" (1994a, p. 142). The problems that causes might be just as much a part of the science of politics, as they affect non-scientists responses to the science that the IPCC presents with respect to catastrophic human-caused environment change, but it is also a matter of the politics of science.

Hulme and Mahony noted concerns about the geographical bias in the selection of Working Group participants that had been raised in a variety of quarters. They quoted former Chair of the IPCC, Bert Bolin, who during his incumbency observed that "right now, many countries, especially developing countries, simply do not trust assessments in which their scientists and policymakers have not participated. Don't you think credibility demands global representation?" They followed this with Kandlikar and Sagar's observation "that a fair and effective climate protection regime that requires cooperation with developing countries, will also require their participation in the underlying research, analysis and assessment." And finished with Lahsen's finding that in interviews "Brazilian climate scientists reflect some distrust of. .. the IPCC, which they describe as dominated by Northern framings of the problems and therefore biased against interpretations and interests of the South" (Hulme and Mahony 2010, p. 709).

## *The Science of Politics and the IPCC*

Those involved in the IPCC face a difficult environment with respect to establishing and maintaining its authority with respect to the science of climate change. They face an equally difficult tack when it comes to establishing and maintaining its authority with respect to governing in conditions deteriorating because of catastrophic human-caused environment change. This distinction is, of course, not as clear or simple as the discussion here might be taken to imply. An overlap between the creation of the two forms of authority that the IPCC requires has already emerged with respect to the geographical bias of IPCC membership. But the IPCC also faces distinct challenges when it comes to establishing authority in the context of the science of politics. The IPCC is, above all else, a political entity. It is, after all, an inter-governmental panel and the politics behind its very creation merit consideration. As do the Summaries for Policy Makers (SPMs) that are contained in IPCC reports. In both cases, the members of the IPCC are being affected by the requirements of politics.

Understanding the IPCC necessitates situating its creation, as it emerged in a particular 'political' context. This context gave rise to a specific form of organization that operated in specific ways. The principal feature of the environment in which the IPCC was created included, according to Hulme and Mahony, "the loss of cultural and social readings of climate and the reframing of climate as 'global'; the rising power of climate modelling and Earth system science; the rise of global environmental politics during the 1980s; the politics surrounding the end of the cold war; and a new 'green' imperialism in European societies" (2010, p. 707).

But there were more overt political objectives associated with the creation of the IPCC. Agrawala has suggested that the emergence of the IPCC "as the politically favoured means of climate change knowledge assessment owed much to American unease about [United Nations Environment Program] and to their desire to find a means of balancing the advocacy positions of the fossil fuel and environmentalist lobbies in the USA" (Hulme and Mahony 2010, p. 706). While, for Boehmer-Christiansen, the way that the IPCC was formed meant that "governments gained the power to veto participants to the IPCC and influence its brief" (Boehmer-Christiansen 1994b, p. 189).

The IPCC is not only a contested scientific space it is also a contested political space. While, as previously noted, Boehmer-Christiansen saw the creation of the IPCC as an attempt to influence policy by members of the scientific community, the times in which it was created were ones of declining government capacity and resolve to address climate change. Whereas green movements had been effective in raising popular and government awareness of the problem of global warming prior to this period, the situation was different at the end of the 1980s. "[F]alling energy prices, over-supply of energy on world markets, privatization and deregulation" meant that those in government "felt increasingly unable to intervene firmly on behalf of the 'greens'. This failure, and the scientific response to it, exposed the limits of scientific advice and the essentially political nature of the IPCC process"

(Boehmer-Christiansen 1994b, p. 185). Thus, it was "reactions to the energy policy implications of early advice, as well as the political agendas which attached themselves to it, [that] led to the demise of an independent advisory body of scientists and its replacement by the Intergovernmental Panel on Climate Change (IPCC) in 1988" (Boehmer-Christiansen 1994b, p. 185).

The point in its reports at which the origins of the IPCC are most evident is the Summary for Policy Makers (SPM) that authors of reports are charged with producing. According to Stocker and Plattner, report authors engage "in a tightly scheduled three-year process that encompasses multiple rounds of draft production, revision and finalization. A long consensus-finding process is needed, from multistep, worldwide reviews of report drafts to the preparation of a carefully worded summary for policy-makers that requires government approval" (Stocker and Plattner 2014, p. 163).

It is necessary, then, for report authors to consider the need for governmental approval. This leads to the concerns about the (scientifically) conservative nature of IPCC reports discussed above. Even though Carraro and co-authors acknowledged that this causes problem, they also argued that it was a factor in the authority of IPCC reports that would be undermined if a new process was adopted.

> The SPMs are the most widely referenced element of each [Working Group] report, largely because their text is negotiated and formally approved by IPCC-member governments. The politically negotiated nature of the government-approval process at the end of each [Assessment Report] can lead to a dilution of key messages. Yet it would be difficult and possibly undesirable to change the SPM production process; IPCC reports are treated as authoritative partly because governments have formally accepted the SPM text. Instead of trying to change the SPM process and format, increasing the prominence of technical summaries (written solely by lead authors) would be useful. Renaming them 'executive summaries' and engaging expert communicators could make them more accessible to policy-makers and the public. More clearly distinguishing between these two types of summaries could facilitate explicit acknowledgment and acceptance of divergent views that sometimes prevail between scientists and governments (Carraro et al. 2015, p. 35).

This passage was quoted in full because it captures something of the two forms of authority that those who write reports for the IPCC are seeking to establish and maintain and the tensions between those who speak for science and those who speak for government. Carraro and co-authors' recommendation with respect to the provision of 'technical summaries' and 'executive summaries' are an attempt to distinguish the issues concerning authority that arise with respect to the politics of science ('technical summary') from those that arise from the science of politics ('executive summary').

As was discussed earlier, one of the crucial questions that arise with respect to the IPCC concerns how it establishes (and fails to establish) authority when it comes to catastrophic human-caused environment change. A variety of concerns with respect to the IPCC arise when it comes to the politics of science. The first concern the preeminence of the IPCC when it comes to scientific accounts of climate change. After that arise various issues for the IPCC associated with the scientists chosen to participate in its Working Groups, with that of geographical bias being one of the more important. For geographical bias can be understood to be a factor in the resistance on the parts of those in government around the world with respect to taking up the IPCC's recommendations for addressing the causes of catastrophic human-caused environment change.

For even after creating an inter-governmental body to provide information and advice concerning climate change, those in government remain reluctant to follow any advice contained in IPCC reports. And this suggests, though bias may be present, that the science of politics is more important in this context than the politics of science. Though note must be taken of the fact that, as Boehmer-Christiansen pointed out, while "energy politics rather than uncertain science have had the decisive impact on global warming policy… this process has benefitted [both] the institutions of big science and the political North..." (1994b, p. 185).

## Conclusion

The issues that arise with respect to the IPCC serve to illustrate both the effects of the politics of science and that of the science of politics. The question of who speaks for science and, in particular, who speaks for climate science is one of the more difficult matters that arise in this context. Constructions of an ideal scientist, who can be argued to have particular cultural attributes, resulted in over-stated possibilities when it came (or comes) to what human could claim to know as true. Ignoring the politics of science may have enabled scientists to continue their work, but it has left them vulnerable when the authority of leading scientists to speak the Truth is contested. At best, the politics of science can be accepted and provided for, but it cannot be ignored (and it will return in this book whenever the climate denial industry is discussed).

The picture becomes even more difficult when the politics of science meets the science of politics in the development of modern, ultimately neoliberal, governance. This new mode of governing, which reconstructed the relationship between the governors and the governed, changed the obligations of governors toward the governed as well as the obligations of the governed to themselves. Even positing obligations to 'self' heralds a new self-understanding amongst those unaccustomed to a self-conception as individuals. Self-interest becomes not only the means to understand but, more importantly, as a means to organize human behavior. The role of government becomes a science of preserving markets and one of dissuading projects of social reconstruction that would reflect an old-fashioned, and disruptive, enthusiasm.

## References

Ablain, M., Legeais, J. F., Prandi, P., Marcos, M., Fenoglio-Marc, L., Dieng, H. B., … & Cazenave, A. (2017). Satellite altimetry-based sea level at global and regional scales. In A. Cazenave, N. Champollion, F. Paul, & J. Benveniste (Eds.), *Integrative study of the mean sea level and its components* (Vol. 58). Space Sciences Series of ISSI. Cham: Springer. https://doi.org/10.1007/978-3-319-56490-6_2.

Akerlof, G. A., & Michaillat, P. (2018). Persistence of false paradigms in low-power sciences. *Proceedings of the National Academy of Sciences*, *115*(52), 13228–13233. www.pnas.org/cgi/doi/10.1073/pnas.1816454115.

Bauer, H. (2017). Science Is Not What You Think. Jefferson: McFarland & Company.

Benestad, R. E., Nuccitelli, D., Lewandowsky, S., Hayhoe, K., Hygen, H. O., Van Dorland, R., & Cook, J. (2016). Learning from mistakes in climate research. *Theoretical and Applied Climatology*, *126*(3–4), 699–703. https://doi.org/10.1007/s00704-015-1597-5.

Boehmer-Christiansen, S. (1994a). Global climate protection policy: the limits of scientific advice: Part 1. *Global environmental change*, *4*(2), 140–159. https://doi.org/10.1016/0959-3780(94)90049-3.

Boehmer-Christiansen, S. (1994b). Global climate protection policy: the limits of scientific advice: Part 2. *Global environmental change*, *4*(3), 185–200. https://doi.org/10.1016/0959-3780(94)90002-7.

Carraro, C., Edenhofer, O., Flachsland, C., Kolstad, C., Stavins, R., & Stowe, R. (2015). The IPCC at a crossroads: Opportunities for reform. *Science*, *350*(6256), 34–35. https://doi.org/10.1126/science.aac4419

Christoff, P. (2016). The promissory note: COP 21 and the Paris Climate Agreement. *Environmental Politics*, *25*(5), 765–787. https://doi.org/10.1080/09644016.2016.1191818.

Cook, J., Nuccitelli, D., Green, S. A., Richardson, M., Winkler, B., Painting, R., … & Skuce, A. (2013). Quantifying the consensus on anthropogenic global warming in the scientific literature. *Environmental Research Letters*, 8(2), 024024. https://doi.org/10.1088/1748-9326/8/2/024024.

Curtis, B. (2002). Foucault on governmentality and population: The impossible discovery. *Canadian journal of sociology*, *27*(4), 505–535. https://doi.org/10.2307/3341588.

Eulau, H. (1969). Tradition and innovation: On the tension between ancient and modern ways in the study of politics. In H. Eulau (Ed.), Behavioralism in political science, (pp. 1–21). New York, NY: Atherton Press.

Farr, J. (1988). Political science and the enlightenment of enthusiasm. *American Political Science Review*, *82*(1), 51–69. https://doi.org/10.2307/1958058.

Foucault, M. (2008). The birth of biopolitics – lectures at the Collège de France 1978–1979. Basingstoke: Palgrave Macmillan.

Herschel, J. (1851). *Preliminary discourse on natural philosophy*. London: Longman, Brown, Green & Longmans.

Hobbes, T. (1946). *Leviathan*, M. Oakeshott (Ed.), Oxford: Blackwell.

Hulme, M., & Mahony, M. (2010). Climate change: What do we know about the IPCC?. *Progress in Physical Geography*, *34*(5), 705–718. https://doi.org/10.1177/0309133310373719.

Kochiras, H. (2009). Gravity and Newton's substance counting problem. *Studies in History and Philosophy of Science Part A*, *40*(3), 267–280. https://doi.org/10.1016/j.shpsa.2009.07.003.

King, M. T. (2001). Security, scale, form, and function: The search for truth and the exclusion of evidence in adversarial and inquisitorial justice Systems. *Int'l Legal Persp.*, *12*, 185.

Kuhn, T. S. (2012). *The structure of scientific revolutions*. University of Chicago press.

Le Quéré, C., Andrew, R. M., Friedlingstein, P., Sitch, S., Hauck, J., Pongratz, J., ... & Arneth, A. (2018). Global carbon budget 2018. *Earth System Science Data (Online)*, *10*(4). https://doi.org/10.5194/essd-10-2141-2018.

Mäki, U. (1995). Diagnosing McCloskey. *Journal of Economic Literature*, *33*(3), 1300–1318.

Mansfield, H. C. (1981). Machiavelli's political science. *American Political Science Review*, *75*(2), 293–305. https://doi.org/10.2307/1961365.

Martin, D. A. (1990). Economics as ideology: on making "The Invisible Hand" invisible. *Review of Social Economy*, *48*(3), 272–287.

Nachmany, M., & Setzer, J. (2018). Policy brief Global trends in climate change legislation and litigation: 2018 snapshot. *London: Grantham Research Institute on Climate Change and the Environment*. http://www.lse.ac.uk/GranthamInstitute/publication/global-trends-in-climate-change-legislation-and-litigation-2018-snapshot/. Accessed 15 June 2019.

Natarajan, V. (2008). What Einstein meant when he said "God does not play dice…". *Resonance, 13*(7), 655–661. https://doi.org/10.1007/s12045-008-0072-3.
Nerem, R. S., Beckley, B. D., Fasullo, J. T., Hamlington, B. D., Masters, D., & Mitchum, G. T. (2018). Climate-change–driven accelerated sea-level rise detected in the altimeter era. *Proceedings of the National Academy of Sciences, 115*(9), 2022–2025. https://doi.org/10.1073/pnas.1717312115.
Nowakowski, A. & Teich, M. (2015, Mar 11). Why Have People Lost Faith In Science? https://www.wuwm.com/post/why-have-people-lost-faith-science#stream/0. Accessed 15 June 2019.
Oreskes, N., & Conway, E. M. (2011). *Merchants of doubt: How a handful of scientists obscured the truth on issues from tobacco smoke to global warming*. New York: Bloomsbury.
Orman, T. F. (2016). "Paradigm" as a Central Concept in Thomas Kuhn's Thought. *International Journal of Humanities and Social Science, 6*(10), 47–52. www.ijhssnet.com/journals/Vol_6_No_10_October_2016/8.pdf. .
Pew Research Center. (2017 December 8). Mixed Messages about Public Trust in Science, Pew Research Center: Science & Society http://www.pewresearch.org/science/2017/12/08/mixed-messages-about-public-trust-in-science/. Accessed 15 June 2019.
Pew Research Centre. (2015, July 23). An Elaboration of AAAS Scientists' Views. http://www.pewinternet.org/2015/07/23/an-elaboration-of-aaas-scientists-views/-fn-13900-3. Accessed 15 June 2019.
Porter, A. J., Kuhn, T. R., & Nerlich, B. (2018). Organizing authority in the climate change debate: IPCC controversies and the management of dialectical tensions. *Organization Studies, 39*(7), 873–898. https://doi.org/10.1177/0170840617707999.
Robinson, V. N. E. (2018). The Origins of General Relativity and the Fabric of Space–Time. https://www.researchgate.net/publication/325988703_The_Origins_of_General_Relativity_and_the_Fabric_of_Space-Time?_sg=xWpP8TcfEChdz62nsXp_wvYz_U1-Hq_X8eUuoIbixOaSbGXhmPrITsni5Ixa9ULmXqW9X3C7Mh2zQWrNKCiwM-yucFDAAEQPl7_JITSP.oDX8tdO2CGRnFKsHCCM9UYltAmkVpjc5ZyAGcc7wCIEIJFXfVf1XnhP0bZbj3s_nIqhzO0KXW1-OQwibwKupaw.
Romm, J. (2015). *Climate change: What everyone needs to know*. Oxford University Press.
Russell, B. (1967). *The impact of science on society*. Routledge.
Saha, A. (2015). The war on climate change: Hubris to realism. *The Fletcher Forum of World Affairs*, 39(2): 49–63. https://www.fletcherforum.org/s/39-2_11Saha.pdf. Accessed 15 June 019.
Scherer, G. (2012). Climate science predictions prove too conservative. *Scientific American, 6*. https://www.scientificamerican.com/article/climate-science-predictions-prove-too-conservative/.
Schiebinger, L. (1987). The history and philosophy of women in science: A review essay. *Signs: Journal of Women in Culture and Society, 12*(2), 305–332. https://doi.org/10.1086/494323.
Schneider, S. H. (2001). A constructive deconstruction of deconstructionists: a response to Demeritt. *Annals of the Association of American Geographers, 91*(2), 338–344. https://doi.org/10.1111/0004-5608.00246.
Stiglitz, J. E. (2015). Overcoming the Copenhagen failure with flexible commitments. *Economics of Energy & Environmental Policy, 4*(2), 29–36. https://doi.org/10.5547/2160-5890.4.2.jsti.
Stocker, T. F., & Plattner, G. K. (2014). Climate policy: Rethink IPCC reports. *Nature News, 513*(7517), 163. https://doi.org/10.1038/513163a.
Suvin, D. (2010). On the horizons of epistemology and science: To Gene Gendlin. *Critical Quarterly, 52*(1), 68–101. https://doi.org/10.1111/j.1467-8705.2010.01924.x.
White, P. (2016). The man of science. In B. Lightman, (Ed.). *A companion to the history of science* (153–163). Chichester: John Wiley & Sons. https://doi.org/10.1002/9781118620762.ch7.
Wiens, J. A. (2016). *Ecological Challenges and Conservation Conundrums: Essays and Reflections for a Changing World.* John Wiley & Sons.
Wray, K.B. (2018). A note on measuring normal science. Scientometrics 117(1) pp 647–650. https://doi.org/10.1007/s11192-018-2870-z.

# Chapter 3
# Ideology and the Inevitability of the Final One Hundred Years of Humanity

Those in government have played and will play crucial roles in the maintenance of a trajectory that leads to the final hundred years of humanity. Those in positions of political power will be blamed for inaction with respect to addressing catastrophic human-caused environment change early enough or fully enough. And they deserve much of the blame, however, though scientists bear some responsibility but are likely to avoid most of the blame. Those in government do not deserve all the blame and they ought not shoulder full responsibility for the maintenance of the trajectory toward humanity's final hundred years. For many members of the communities they governed supported and will continue to support the ongoing failure to adequately address the causes of catastrophic human-caused environment change. The most important of those who support governmental inaction are those involved in the climate change denial industry; so important that they deserve a separate chapter.

Support for the continuation of a trajectory leading to humanity's final hundred years has not and will not be in the form of widespread demands to ignore mainstream climate science and resist agendas of radical social and economic change. Nor will it come in the form of public movements calling for the maintenance of a trajectory that leads to the final hundred years of humanity. Rather the maintenance of this trajectory will result, for the most part, from the intellectual difficulties associated with responding to the causes of catastrophic human-caused environment change and from ideas concerning what is normal, natural, desirable and possible that condition and often constrain the demands for action that are directed toward those in positions of political power.

The affects discussed in this chapter are often taken to be ideological. They condition and constrain behavior in subtle ways that result in an ongoing failure to do enough to address the human causes of catastrophic environment change. And they do not lead to ignoring or dismissing mainstream climate science, which is part of climate change denial, but of failing to adequately respond to this science because an adequate response requires fundamental or radical (at the root) change. Preventing radical change (i.e. constraining enthusiasm) might well be taken to be the principal role of ideology. "'Ideology' is the name for this discursive response that both



I. Cook, *The Politics of the Final Hundred Years of Humanity (2030-2130)*,
https://doi.org/10.1007/978-981-15-1259-9_3

acknowledges the wastefulness of capitalism… and attempts, nevertheless, to maintain capitalism in more or less its current state" (Thompson 2012, p. 896).

Ideas concerning the normal, the natural, the desirable and the possible represent key elements of political belief systems. These systems result from "the interrelationships of attitudes and beliefs relevant to politics… as a network of interacting nodes. … The belief system then consists of all of the relevant nodes and their connections with one another". This system is comprised of "operational components (i.e., positions on specific issues, such as government spending)" and "symbolic components (i.e., affective attachments to political groups and labels, such as ideological identification or party identification)" (Brandt et al. 2019, p., 1). Symbolic components are particularly important in this context, as people use them to confirm their position as one shared by others. As Mason put it, "ideology … is both a system of beliefs… [and] a sense of connection to like-minded others (even when this is not true in terms of actual opinions)" (Mason 2018, p. 867).

Resistance to policies that would result in effective action to address the human causes of catastrophic environment change reflects both the effects of specific belief systems and identification with specific others as 'like-minded'. So, the ideological effects discussed in this chapter are magnified by the fact that they function within communities of like-minded people who, because group membership is crucial for identity and sound argument is not, cannot be moved by argument. This can lead people to support certain political positions because of the others who articulate those positions and not because of what they themselves believe. For, "when it comes to political belief systems, our symbolic attachments to parties and labels are a more important part of the belief system than the actual policy positions" (Brandt et al. 2019, p. 9). The result, according to Mason, is that "the 'otherness' of ideological opponents, more than issue-based disagreement, … drives liberal-versus-conservative rancor" (Mason 2018, p. 867).

And as symbolic attachments outweigh policy commitments, conversation with those with whom we do not identify, those who do not belong to our in-group, becomes impossible. For Mason, "social distancing … is occurring without much regard to actual issue-based disagreement." People divide because "of their identity-related feelings about who is 'in' and who is 'out'" and this leads to a commitment to my in-group, or 'team', becoming more important than a desire to achieve an outcome. The problem is that "team names without issue knowledge can generate political conflict that is unmoored from distinct policy goals. This is likely to lead to a less compromise-oriented electorate. After all, if policy outcomes are less important than team victory, a policy compromise is a useless concession to the enemy" (Mason 2018, p. 884).

The result of the in-group effect is that most voters are voting for team (party) and not policy. People are not in favor of exacerbating the human causes of catastrophic environment change. They simply identify with parties that offer policies that do not address the human causes of catastrophic environment change because they do what political institutions must be expected to do: maintain the status quo. And this is unlikely to bother most voters, who are committed to the status quo and not an ideology. As Kinder and Kalmoe wrote, "genuine ideological identification – and binding dispositional commitment to an ideological point of view – turns out to be rare" (Kinder and Kalmoe 2017, p. 7). For them, group membership determines ideology, rather than ideology determining group membership and "public opinion arises… primarily from antipathies and attachments of group life" (Kinder and Kalmoe 2017, p. 136).

This means that, for example, "American elections are, first and foremost, affirmations of loyalty to party, and campaigns are, first and foremost, efforts on the parts of parties to reinforce and activate their supporters" (Kinder and Kalmoe 2017, p. 137). It is the fact that the parties that voters support are implicated and integrated in political processes that reproduce the status quo that means that not enough is done to address catastrophic human-caused environment change and avert a trajectory toward the final hundred years of humanity.

This is not to suggest that the major parties in the US mount the same mode of defense of the status quo. The Republicans have moved to implacable resistance to action to address the human causes of climate change and taken their supporters with them. "Republican supporters were once as likely to believe climate science as Democrats" (Merkley and Stecula 2018, p. 260). Democrat leaders have differentiated themselves from their Republican counterparts in accepting that the Earth's climate is changing because of human action and their followers line up with them in believing in the reality of catastrophic human-caused environment change.

Though Representative Alexandria Ocasio-Cortez and Senator Ed Markey introduced a resolution outlining the principal features of a Green New Deal resolution in Congress in March 2019, there is little to suggest that all the Democrats in Congress will support it. Within 2 weeks of the introduction of the resolution a coalition of "centrist Democrats" in Congress declared that they would "kill the Green New Deal". Utah representative Ben McAdams led this coalition, which was "pushing back on calls for a Green New Deal – the only proposal on the table that remotely matches the ambition science demands" (Aronoff 2019).

The greater importance of party over policy for voters is not limited to the US, according to Kinder and Kalmoe, and what they "found in United States matches up well with others have reported from other places" (Kinder and Kalmoe 2017, p. 131). Politics is conducted, then, in the context of tensions around identity and group membership and not through tensions about the way society should be (such as being capable of sustaining humanity on the parts of its members). Maintaining a sense of self requires maintaining a connection to party and other party supporters, neither of which is likely to risk the maintenance of this connection by introducing policies that disturb people's connections. Switching to policies that require fundamental change threatens stability of identity and requires effort to address.

Introducing issues associated with identity, though, should not be taken to be an interest in demography and the personal characteristics of those who, for one reason or another, see no need to address the human causes of catastrophic environment change. Hornsey and colleague's meta-analysis of studies on this matter makes clear that "traditional societal faultlines of gender, age, sex, race, and income seem to be of little relevance in determining levels of climate change skepticism" (2018, p. 625). The studies Hornsey and co-authors analyzed also indicated limited effects for "education, (subjective) knowledge, and experience with extreme weather events" (2018, p. 626). Their conclusion from the research they studied was that "these intuitively appealing determinants of climate change belief were overshadowed in predictive power by values, ideologies and political affiliation" (Hornsey et al. 2018, p. 626).

Social distancing is important for preventing conversation, especially conversation concerning meaningful change and action to address the human causes of catastrophic environment change. A variety of other ideological affects, however, mean that those in positions of political power will receive support for ongoing inaction that means that a trajectory toward the final hundred years of humanity will be maintained. One of the most important of these ideological affects is what has been labeled the 'climate change denial industry'. The role of this industry is sufficiently important in the maintenance of a trajectory leading to the final hundred years of humanity that the next chapter of this book has been devoted to a discussion of the industry and the likely strategies of members of this community during the early period of the final hundred years of humanity.

As the climate change denial industry is the subject of the following chapter, here the focus will be on other ideological factors that support the maintenance of a trajectory toward the final hundred years of humanity. Most of these ideological factors operate predominantly in western countries. Some operate in other countries of the Global North. One functions only in countries of the Global South. So, most of these factors produce significant effects on those in western countries, but because of globalization, are increasingly affecting people in non-western societies. Some factors are present in the minds of significant numbers of people. Others affect fewer people but are important because of the people they affect.

The theme of this chapter is that those in positions of political power have not introduced the changes necessary to divert from a trajectory toward the final hundred years of humanity because of a lack of support for or resistance to those changes on the parts of the people they govern. These factors are most likely to apply in representative democracies, in which the people's will is a more powerful control on political leaders than in non-democratic political systems. But many will apply to people non-democratic systems.

Seven (ideological) factors that lead people to resist, or fail to support, the changes necessary to avert the continuation of a trajectory toward the final hundred years of humanity, will be discussed in this chapter. The most basic of these is most people's resistance to significant change. Many people can tolerate and even enjoy some change, but fundamental change presents too great a challenge for most people. Another reason to resist or fail to support the implementation of the changes necessary to avert a trajectory toward the final hundred years of humanity is that it requires coming to terms with a problem that, due to its causes and consequences, is very hard to comprehend.

But it is not just hard to imagine, it is also an uncomfortable problem for those in Western countries, who Ariès (1981) suggests are trained to hide death and to hide from death. And even if people can comprehend the problem, contemplating the scale of what must be done to address catastrophic human-caused environment change is a daunting, and for most of us overwhelming, task. For many people, then, it is easier not to think about the problem or not to think about it for very long or with great care. And it is decidedly easier not to think about what must be done to address the human causes of catastrophic environment change and avert a trajectory toward and into the final hundred years of humanity.

And if addressing the human causes of catastrophic environment change requires the end of capitalism, as it seems it does, then additional resistances arise. Capitalism is normal or even natural for those raised in Western societies, and for increasing numbers in other societies, as it provides for the selfish individualism that they see as basic to human being. They will resist or be unwilling to support a change from a capitalist system in which normal selfishness is not available. Further, many of those in capitalist societies see those societies as providing for their needs through maximizing opportunities for consumption (for some authors, they practice an ideology of consumerism). And this desire for consumption can go further for them than satisfying material needs, as 'retail therapy' is increasingly accepted and practiced as a means for addressing emotional (rather than material) needs through consumption. This means that people keep buying and consuming even when they do not truly need the items they buy and consume.

While everyone in capitalist societies sees selfish individualism as normal, only some of them have been raised to expect that the world will end. For these people a trajectory leading to the final years of humanity is anticipated, and environmental catastrophe fits that for which they have been taught to prepare. Even though this affects only some, these people are differently oriented toward the final hundred years of humanity from other members of their societies. Importantly, many of the people who believe in end times are Christians or Muslims and their numbers will grow as those religions continue to spread and the signs that humanity is approaching its final hundred years become ever clearer.

Another factor, which only affects some people, is a reluctance to cooperate in producing and implementing international agreements to address catastrophic human-caused environment change on the parts of those in government in countries of the Global South. People in countries of the Global South, including those in positions of political power, have legitimate reasons to resist calls for them to address the human causes of catastrophic environment change because they believe that the problem has been and continues to be the principal responsibility of those in countries of the Global North. As a result, they should be addressed by those in countries of the Global North. Further, some argue that people in countries of the Global South must be allowed to continue to pursue economic development that contributes to the human causes of catastrophic environment change while those in countries of the Global North work to avert a trajectory toward the final hundred years of humanity.

Each of these factors is discussed in a separate section below. No one of them is enough to explain government inaction with respect to addressing the human causes of catastrophic environment change. Each contributes to that inaction, however, and helps to explain why a problem can be recognized to exist yet nothing is done to address that problem. There will come a point at which meaningful, indeed drastic action, is taken to avert a trajectory leading to the final hundred years of humanity. Indeed, this is addressed in Chapter Six, where the politics of two desperate measures (radical population reduction and stratospheric aerosol injection) are discussed. The seven ideological factors discussed below will delay action, though, to the point that these desperate measures are likely to prove insufficient to avert the end of humanity.

Given that many form parts of a belief system, these factors are often interrelated and sustain each other (though this effect ought not to be over-stated, as in-group effects allow the maintenance of contradictory values). Hornsey and co-authors wrote "that people's perceptions of risk are influenced by their concept of how society should be structured, and that this conceptualization leads them to uphold specific cultural values" (Hornsey et al. 2018, p. 623). An important finding of research into responses to information about catastrophic human caused environment change is that "people who subscribe to relatively individualistic and hierarchical values are more inclined to value elites and the status quo, and so are motivated to disbelieve that industry poses a risk to the environment" (Hornsey et al. 2018, p. 623).

Seven 'ideological' reasons that explain insufficient support for government action to address the human causes of catastrophic environment change will be discussed in this chapter. These are: because the fundamental or radical changes required to address catastrophic human-caused environment change are too great for people to accept (conservatism); because the problem of catastrophic human-caused environment change leading to the final hundred years of humanity is too hard to conceive (scope); because the change required to address the problem of catastrophic human-caused environment change is too much to conceive (scale); because capitalism seems normal or natural to them and political action to address catastrophic human-caused environment change is not in their individual interest (selfish individualism); because political action to address catastrophic human-caused environment change threatens the production-projection of identity through consumption (consumerism); because they have been brought up to believe in the end of the world, so catastrophic human-caused environment change is not a problem for those in positions of political power (apocalypticism); and, seventh, because people in countries of the Global South see those in countries of the Global North as responsible for catastrophic human-caused environment change (postcolonialism).

## Because the Fundamental or Radical Changes Required to Address Catastrophic Human-Caused Environment Change Are Too Great for People to Accept (Conservatism)

Significant change, for most people, is uncomfortable and, as a result, undesirable. Doing what we have always done takes less effort than changing what we do. One way to understand this is through the (albeit) contested concept of 'ego depletion' (Dang et al. 2019). "The core idea behind ego depletion is that the self's acts of volition draw on some limited resource, akin to strength or energy and that, therefore, one act of volition will have a detrimental impact on subsequent volition" (Baumeister et al. 2018, p. 1252).

Every choice we make takes work and reduces our energy level. It requires effort, and this is always experienced physiologically in terms of loss. The more acts are automatic and done unreflexively the more energy is conserved. People seek to preserve their energy by repeating past behavior. Simply doing what we have always done does not draw upon this depletable psychological 'resource', this energy. Taking conscious, or active, control does. As Baumeister and colleagues put it, "responding in a controlled (as opposed to automatic) fashion would cause ego depletion … Hence, staying in the automatic realm would help conserve this resource" (Baumeister et al. 2018, p. 1263).

Baumeister and co-authors found it relatively easy to cause ego depletion in their subjects and inferred from this that "the extent of the resource is quite limited, which implies that it would be seriously inadequate for directing all of a person's behavior, so conscious, free choice must remain at best restricted to a very small proportion of human behavior." The result, in their view, is that "most behavior would have to be automatic instead of controlled" (Baumeister et al. 2018, p. 1263). They argued that the results from the four studies that they conducted, which demonstrated various effects of ego depletion, pointed "to a potentially serious constraint on the human capacity for control (including self-control) and deliberate decision making" (Baumeister et al. 2018, p. 1264).

Another way to explain resistance to change is through the concept of 'cognitive consistency'. Here again, though, the effort and energy associated with making change is important. Cognitive consistency is particularly significant, as Jost suggested, in the context of "ideology, which can be thought of as a network of interrelated beliefs, values, and opinions. People are highly resistant to persuasion when it comes to a belief that is ideologically relevant, because if they change one belief they are obliged to reconsider other beliefs that are logically dependent on the initial belief…" (2015, p. 613). Belief networks hold constellations of belief in place. The elements of these constellations support and maintain each other.

The most resistant to change, of course, are those whose values are compatible with an existing social order. Thus, Jost found that those who endorsed "system-justifying statements are less concerned about environmental problems, more skeptical about climate change, and less likely to report engaging in recycling and other ecologically minded behaviors. (Jost 2015, p. 615). Along related lines, Lacroix and Gifford observed that "individualists typically are dismissive of environmental and technological risks because they usually want to avoid restrictions on commerce and industry." They also found that people "who hold hierarchical cultural beliefs tend to be dismissive of environmental and technological risks because they threaten the authority or competence of those in power" (2018, p. 752).

Pointing to research that "supported the notion that conservatism opposes the specific solutions generally advocated by environmentalist organizations because they threaten the status quo" (2018, 164), Panno and Carrus seek better way to address conservatives. They recommended presenting "the change brought about by new regulations as endorsing the system…" and crafting "proenvironmental policies so as to appear consistent with an economically centered positive attitude towards inequality" (Panno et al. 2019, p. 164). They even look for a way to represent action to address the human causes of catastrophic environment change in terms of "a frankly conservative-populist outlook, one emphasizing the value of tradition, the value of conservation, the perils of modernity and materialism, and the wisdom of the old ways" (Panno et al. 2019, p. 165).

It may be possible to follow Panno and Carrus's advice and seek to make action to address the human causes of catastrophic environment change appear not to threaten the status quo. But there are limits to how long the illusion that fundamental or radical change is unnecessary can be maintained. Averting a trajectory that leads to the final hundred years of humanity means billions of people living differently. People must change. This necessitates that they move away from the automatic behaviors that are important to the continuation of a trajectory toward the final years of humanity. It also requires changing beliefs that will have a cascading, or domino, effect and produce changes in other beliefs. All of this is too much for most, perhaps all, people. The result is that most people will be likely to continue to support their government's failure to adequately address the human causes of catastrophic environment change for which the members of their society are responsible.

## Because the Problem of Catastrophic Human-Caused Environment Change Leading to the Final Hundred Years of Humanity is too Hard to Conceive (Scope)

A related explanation for a lack of popular support for effective government action to address the human causes of catastrophic environment change is that people cannot and do not want to fully come to terms with the problem. Two factors may be at work here. The first, and more important, is that the problem is simply very difficult to comprehend. The human mind has limits and catastrophic human caused environment change exceeds those limits; or, at least, exceeds those limits for most people. A second factor, though specific to western cultures, is that this subject raises the specter of death, which is one from which westerners have learnt to hide.

As Brick and van der Linden argued, "climate change is psychologically unprecedented as a social dilemma because of the unique cognitive and social challenges it presents to people" (Brick and van der Linden 2018, p. 32). As they suggested, "it is very difficult for humans to grasp threats that are sometimes perceived as invisible, gradual, distributed and long-term" (Brick and van der Linden 2018, p. 32). The problem with catastrophic human-caused environment change is that it does not trigger the right response in brains unequipped to respond to this type of threat; for "abstract and invisible threats aren't terrifying to brains that evolved to solve local, experiential and imminent problems" (Brick and van der Linden 2018, p. 32). Catastrophic human-caused environment change, then, is the name for "an abstract statistical phenomenon that does not easily trigger our intuitive, associative and affective warning system, which is largely based on experience and evolved to map visible environmental cues onto concrete threats" (Brick and van der Linden 2018, p. 34).

This might be thought to be unrelated, but most, if not all, westerners' inability to engage with death in a healthy manner is another factor that can be argued to be at work in making catastrophic human-caused environment change hard to address. As Ariès pointed out many years ago (1981), western culture has erased or hidden death and produced a collective refusal to engage with it. Western culture required that death not come to mind, Ariès argued, because it had become fully oriented to material satisfaction (a conception of human existence demanded by utilitarian/ Benthamite felicific calculus) and was devoted to the pursuit of happiness.

According to Ariès, "around the middle of the twentieth century in the most individualistic and middle-class parts of the West… [arose] a conviction that the public demonstration of mourning, as well as its too-insistent or too-long private expression, is inherently morbid. Weeping is synonymous with hysteria. Mourning is a malady" (Ariès 1981, p. 580). In this "new sentiment characteristic of modernity: one must avoid… for society's sake…the disturbance and the overly strong and unbearable emotion caused by the ugliness of dying and by the very presence of death in the midst of a happy life" (Ariès 1981, p. 87).

People in these societies acquired "the moral duty and the social obligation to contribute to the collective happiness by avoiding any cause for sadness or boredom, by appearing to be always happy, even if in the depths of despair" (Ariès 1981, p. 93–4). The result is that the dying are not simply secreted in hospitals and their bodies kept from view. Nor is mourning merely refused. "The denial of death has gone beyond the bereaved and the expression of mourning. It has extended to everything that has to do with death…" (Ariès 1981, p. 580).

So even it was not a massive slowly developing problem operating at a distance, the fact that catastrophic human-caused environment change requires contemplating death, and death on a large scale, makes it difficult for many, especially in the West, to engage with at an intellectual level. And if the problem is hard to contemplate and uncomfortable to say with, then it is unlikely that it will take up a crucial position on the political agenda to which voters and, as a result, their governments respond. Lingering senses that there may be a problem and that someone ought to be addressing that problem may waft about somewhere in the social ether. But it will not crystallize into a demand for immediate and radical action on the parts of those in positions of political power.

## Because the Change Required to Address the Problem of Catastrophic Human-Caused Environment Change is Too Much to Conceive (Scale)

While the problem of catastrophic human-caused environment change is close to impossible to comprehend, the solutions to that problem are equally difficult to imagine. Farenhold captured a number of the sources of popular support for government inaction to address catastrophic human-caused environment change in the following quotation and identified much of the difficulty that arises with respect to the ideological affect discussed here: "It is a global problem, with no obvious villains and no one-step solutions, whose worst effects seem as if they'll befall somebody else at some other time. In short, if someone set out to draw up a problem that people would not care about, one expert on human behavior said, it would look exactly like climate change" (Fahrenthold 2009, p. 1). That there is no simple cause to address and no single action to take are important to people feeling that there is nothing that can be done. "Naomi Klein and others have argued that it's not that people don't care: it's just that the problem is so enormous and encompassing that people feel helpless and disempowered" (Brick and van der Linden 2018, p. 30).

Even those who accept the existence of a problem that they cannot comprehend will struggle to understand and do what needs to be done. These people want to do something. Nothing they can think to do seems to have anything much to do with truly addressing the problem, however, which makes their actions seem, at best, pathetic. For Latour, "there is a form of practical climatoscepticism very different from the cognitive one." As a critic of the conception of science as source of absolute Truth, Latour can only position himself "behind the I.P.C.C. report (not the same thing as 'believing' in it)." Yet he feels "very much that I am a skeptic since I don't know what to do about it, apart from a few pathetic gestures like sorting my rubbish and limiting my carbon footprint (and feeling guilty about it)" (Latour 2014, p. 54).

There may well be more at work than simply the overwhelming nature of the problem of addressing catastrophic human-caused environment change, however. For many authors, neoliberal, anti-state, anti-collectivist ideology has rendered people incapable of conceiving and engaging in the collective projects necessary for diverting from a trajectory toward the final hundred years of humanity. As Thompson put it, "the subject lacks the coordinates within globalized capitalism with which to orient him- or herself towards a socialized good" (2012, p. 900). The result is that "we now 'think locally, [and] act locally', either because we cannot situate ourselves in terms of the global, or because the global, having actually become global, seems to operate so far beyond what we perceive as our sphere of influence that we pragmatically choose to ignore it" (Thompson 2012, p. 900–1).

People may want something to be done to address catastrophic human-caused environment change but not only do they not know what they want done, they no longer believe in the collective action necessary to address this problem. Instead they make the sorts of "pathetic gestures" that Latour felt he made. Or, as Thompson put it, "we make ourselves busy micro-ethically. And, as we might expect in an economic system absolutely dependent on growth and consumption, the most widespread mode of microethics today is the effort to consume ethically as individuals" (Thompson 2012, p. 901).

The obvious problem with this response, as Maniates pointed out, is that it results in "the individualization of responsibility. When responsibility for environmental problems is individualized, there is little room to ponder institutions, the nature and exercise of political power, or ways of collectively changing the distribution of power and influence in society…" (Maniates 2002, p. 45). The less obvious problem is that people do not engage in collective action through public institutions to address the human causes of catastrophic environment change. "Instead, the serious work of confronting the threatening socioenvironmental processes… falls to individuals, acting alone, usually as consumers" (Maniates 2002, p. 45).

This effect is obviously related to the difficulties associated with contemplating the problem. Indeed, the quotation from Fahrenthold with which this section began makes clear that the problems are related. But if those to whom they are supposed to respond do not have a clear sense of what needs to be done and how it can be done they cannot place pressure on those in positions of political power to undertake effective programs to address catastrophic human-caused environment change. This is not to suggest that individuals and companies are not proposing their favored solutions for some of the causes of catastrophic environment change. Indeed, this contributes to inaction on the parts of those in positions of political power because it means that any conversations around addressing catastrophic human-caused environment change that might occur are rendered difficult through the conflicting interests of those participating in those conversations.

## Because Capitalism Seems Normal or Natural to Them and Political Action to Address Catastrophic Human-Caused Environment Change Is Not in Their Individual Interest (Selfish Individualism)

The reduction of the problem of catastrophic human-caused environment change to one of individual action reflects a larger dynamic at work in late capitalist societies. This is the promotion of a selfish individualistic way of life. For Lasch a culture of narcissism has developed in countries of the Global North (2018); whereas for Jacques people in those countries live in an "age of selfishness" (2002). While for Gerhardt, advanced capitalist society is "a selfish society, a society which sustains individualism, greed and materialism at the expense of collective interests and the needs of the social group as a whole" (2010, p. 13).

Selfish individualism has a variety of effects. Its most important in this context are to bring people to measure their quality of life in terms of their personal satisfaction via the accumulation of wealth and possessions and for them to conceive of social progress in terms of an increased capacity to produce objects for consumption. According to Gerhardt, those in the West "take it for granted that human relationships will in many ways be subordinated to our concern with expanding productivity" (Gerhardt 2010, p. 10). The result, in her view, is that "we are living in an impoverished emotional culture, the end product of decades of individualism and consumerism, which have eroded our social bonds" (Gerhardt 2010, p. 12).

Even though debates concerning whether human beings are naturally selfish persist, they are futile to the extent that they posit an either-or position: i.e., humans are always selfish, or they are always selfless. Sonne and Gash offer the more nuanced, and superior, construct of a selfish-selfless spectrum and make clear that where people will find themselves on this spectrum will reflect both genetic and social factors. But as they argued, "the survival and success of large cooperative societies requires most of their members to mute their innate selfish drives and strengthen their selfless behavior." This affect results from "complex combinations of genetic and neuronal factors, including parenting, [that] underlie the spectrum of selfish–selfless behaviors" (Sonne and Gash 2018, p. 3). The consequence of this is that, while people's upbringings would determine where they fell on the selfish-selfless spectrum, this was not permanent and intervening (necessarily social) factors could cause a repositioning on the spectrum.

In emphasizing "the importance of both genetic and neuron-based heredity systems and reward processes in strongly influencing actions toward others and cooperative behavior" (2018, p. 13), Sonne and Gash established that selfish individuals could become significantly less selfish when exposed to the right influences. This reflected research associated with brain or neural plasticity. "Evidence from population groups ranging from incarcerated juveniles, adopted children, twins, and meditators point to the important role of neuroplasticity and reward learning circuitry in forming and reforming of neural connections that determine our behavior" (Sonne and Gash 2018, p. 13). The most important of this evidence showed that "individuals with genes associated with developing dangerous social disorders such as callous, unemotional traits have the potential to modify those conditions using cognitive therapeutic interventions (e.g., strong positive parenting, compassion training) to change where they reside on the social spectrum" (Sonne and Gash 2018, p. 13).

Two points arise from this. The first is that whether people are closer to the selfish end of the selfish-selfless spectrum reflects the interaction between their genes and their social environments. The second is that the obsession with self that characterizes western societies is a result of social processes. For, even though it might be true for most people raised in western societies, according to Fiske, the "Japanese often do not want autonomy, freedom, and individual choice; they want to belong to enveloping, totally encompassing communal sharing groups" (1991, p. 203).

The social processes that position people on the selfish–selfless include those associated with being raised Christian or Muslim, which "emphasized the individual's personal relationship to God, responsibility for salvation, and autonomy" (Sampson 2000, p. 1427). But Protestantism's effects go further, as the "individualistic Protestant ethic even… [plays] a role in psychology's approach to understanding social problems as something more about individuals than the systems they inhabit" (Sampson 2000, p. 1427). As Gerhardt suggested, Adam Smith "could not have imagined the extent to which his concept of 'Homo economicus' driven by self-interest in his business dealings, would come to stand for the whole of human nature and convince many people that a selfish way of life is the only way of life" (Gerhardt 2010, p. 10).

Addressing catastrophic human-caused environment change makes no sense to a selfish individual. There are many ways to characterize the thinking of the selfish individual trained in western societies. But it is well captured in Bentham's characterizaton of the utilitarian rational calculator. This individual assesses all actions in terms of the amounts of pleasure and pain it produces. For, according to Bentham, "Nature has placed [people] under the governance of two sovereign masters, pain and pleasure. It is for them alone to point out what we ought to do…". His "principle of utility is … that principle which approves or disapproves of every action whatsoever according to the tendency it appears to have to augment or diminish the happiness of the party whose interest is in question" (Bentham as quoted in Hirschauer et al. 2015, p. 652).

Decision-making in this understanding isolates an action in order to create a temporal and spatial boundary for the calculation – otherwise it might be endless and must include everything that might bear upon the calculation. An individual calculation of the effects of an isolated action becomes the measure of right and wrong. "According to Bentham, the values of all pleasures and pains are to be added up and balanced against each other. If the balance is on the side of pleasure, the action under consideration will increase individual happiness and vice versa" (Hirschauer et al. 2015, p. 652).

A focus on quantities of pleasure and pain associated with a present action creates immediate problems for collective (political) action to address the human causes of catastrophic environment change. It may even render it impossible. If 'here' is an individual calculation of personal pleasure and pain and 'there' is collective action to address catastrophic human-caused environment change, then we not be able to get 'there' from 'here'. So, this way of thinking leads immediately to a refusal to support government action to avert a trajectory toward the final hundred years of humanity. It makes complete sense, then for Bartlett to lament, in a piece for The Wall Street Journal in 1994, that "every serious study that has ever been done on the subject confirms that the U.S. is paying an enormous cost for environmental regulation" (p. 1).

And he is correct. And every subsequent calculation will come down to the same thing: for a selfish individual focused solely on their interests there are no conditions under which contributing to any action, including to government action to address the human causes of catastrophic environment change, is the right course of action. Above all, they cannot accept losing their jobs to prevent catastrophic human-caused environment change. There is simply no individual calculation that can make sense of this. We cannot blame American miners and other workers who turned to Trump. For he "promised to provide what workers need and want: 'jobs, jobs, jobs!' … [and] committed to scaling back many safety and environmental regulations, which he described as destroying jobs" (Lamont et al. 2017, p. S165).

Selfish people are also very likely to represent selfishness in a positive way and to defend it as not only natural but also necessary. Some even do so in the context of catastrophic human-caused environment change. Predictions of the end of humanity were wrong, according to Bartlett, because they underestimate the capacity of selfishness to solve to the problem. For Bartlett, for example, "the history of [humankind] shows that economic incentives have consistently led to new discoveries of raw materials, changed methods of production, and changes in consumption that have put off the day of reckoning year after year, despite predictions of impending doom" (Bartlett 1994, p. 2).

Two other aspects of selfish individualism merit brief attention here. The first is that selfish individuals are less likely to see group participation as of value in itself and more likely to participate in groups to serve their personal interests. This is, after all, the essence of Hobbes's and most, probably all, of the other social contract theories of the state: For Hobbes, because people "are naturally self-interested, yet… rational, they will choose to submit to the authority of a Sovereign in order to be able to live in a civil society, which is conducive to their own interests" (Friend 2004).

Selfish individualism also makes people more conservative because they either do not participate in groups or do so instrumentally, which makes them less likely to be changed by participation in the group. For research shows that people are more likely to change their views as to the desirability of an existing social and political order when they are members of crowds. In short, and returning to a previous discussion, people are more likely to persist in specific ways of thinking when their thinking is not affected by participation in groups.

Wagoner and colleagues rejected the "claim that crowds are essentially conservative… [as] wrong in its analysis (crowds do contribute to meaningful social change) and also wrong in its premise (that the psychology of crowd members is… essentially resistant to novelty)." Even though most crowds do not produce calls for radical change, they argued, "social change is very often associated with crowd events and crowd behaviour. And it is precisely in crowd events – in particular crowd events in which there is struggle with those in power – where the psychology of participation is recreated in the form of novel understandings of both self and appropriate action" (Wagoner et al., 2018, p. 25).

The more that people in western societies have been conditioned such that they fall close to the selfish end of the selfish-selfless spectrum the less likely they are to demand or even support action to address catastrophic human-caused environment change. It simply makes no sense for them to do so. And if governments in western countries must lead such action, as their historical and ongoing contributions to the problem and present capacities to act suggest, then no such leadership will be called for from the citizens of those countries.

Further, the less people participate in groups, especially crowds, and the more they do so to satisfy their individual interests individually the less likely they are to develop an interest in change - especially change that imposes individual costs, which, for the selfish individual cannot be compensated for by some pay off associated with group membership. Social factors may not be the only ones at work in this context, though, as evidence is growing concerning increased levels of autism caused by the herbicide glyphosate (Roundup) (Good 2018) and maternal stress (Varcin et al. 2017). Irrespective of its cause, though, there can be no doubt that social factors play a key role in the intensifying selfish individualism of those in western countries. This and, seemingly, its growth and increasing intensity amongst people of non-western countries will prevent to many people from demanding and supporting political action to address catastrophic human-caused environment change and result in the continuation of a trajectory toward the final hundred years of humanity.

## Because Political Action to Address Catastrophic Human-Caused Environment Change Threatens the Production-Projection of Identity Through Consumption (Consumerism)

A related reason that most people in western societies do not demand and many do not support political action to effectively address the human causes of catastrophic environment change is that such action is not compatible with their insatiable need to consume. People in western societies are not only brought up to be individuals, they are brought up to share in the ideology of "possessive individualism"

(Macpherson and Cunningham 1962). As Xavier put it, "in our global culture, being a consumer is what ultimately defines the subject" (2018, p. 3). In his view, "consumption has arguably become the main definer of our culture, the chief basis of the social order; rather than being merely an aspect or part of our lives, of our milieu, we all become creatures defined by the 'age of consumption'" (Xavier 2018, p. 2).

Following this line of thought, Sklair wrote of a "culture-ideology of consumerism" and referred to "the transformation of excessively above-subsistence consumption from a sectional practice of the rich throughout human history to a globalizing phenomenon directed at the mass of the population in the era of capitalist globalization" (2012, p. 220). For Sklair, and others, the emergence of the ideology of consumerism was, in part, because "new mass media and communication technologies made it very easy for new consumerist lifestyles to become the dominant motif for these media, which became in time extraordinarily efficient vehicles for the broadcasting of the culture-ideology of consumerism globally" (Sklair 2012, p. 220–1).

As the comment before the preceding quotation suggests, many authors have argued that consumerist ideology is spreading beyond western societies, such that ever more people see any attempt to deny them opportunities to consume as impinging on their personal freedom and expression of self. Kazim, for example, wrote of the emergence of hyper-consumerism in the United Arab Emirates and "its role as a primary means for identity construction" (2018, p. 368).

Other authors note the rise of retail therapy. As most, perhaps all, readers will know, *and that in itself is important*, in retail therapy consumption becomes a means to address negative or unhappy states of mind. As Irwin observed, "retail therapy can be, and frequently is, used to reduce anxiety, stress, sadness, and other negative emotions in low-stakes situations. A third of the United States population, including both men and women, shops as a way to alleviate stress" (2018, p. 14). In 2006, Elliot wrote that "books espousing to help the reader pick the correct 'type' of retail therapy have been published…" (p. 639).

Aware of the extensive criticism of the practice, Irwin defended retail therapy on the grounds that it "can be used intermittently to regain a sense of personal control or provide an escape from a negative environment or headspace… [and] is an effective and valid form of self-care that requires little time or money to reap the benefits" (Irwin 2018, p. 15). In a similar vein, Rick, Pereira and Burson proposed "that retail therapy has been viewed too negatively" and they defended shopping as "an effective way to minimize sadness that lingers (residual sadness) following a sadness-inducing event" (2014, p. 373).

But even if they are not compelled to consume to relieve a psychological disorder, consumers are still governed by the dictates of fashion and will be compelled to consume in order to keep up with the current trends. And these current trends are changing faster than they have ever done and has led to the emergence of the phenomenon of 'fast fashion', which "is a term used to describe the readily available, inexpensively made fashion of today. The word 'fast' describes how quickly retailers can move designs from the catwalk to stores, keeping pace with constant demand for more and different styles" (Bick et al. 2018, p. 1). The crucial attribute of the 'fast fashion' model is that it "encourages consumers to view clothing as disposable" (Bick et al. 2018, p. 1).

According to Bick, Halsey, and Ekenga, fast fashion means that, "globally, 80 billion pieces of new clothing are purchased each year, translating to $1.2 trillion annually for the global fashion industry." The US "consumes more clothing and textiles than any other nation in the world. Approximately 85% of the clothing Americans consume, nearly 3.8 billion pounds [1.72 billion kilograms] annually, is sent to landfills as solid waste, amounting to nearly 80 pounds [36.2 kilograms] per American per year" (Bick et al. 2018, p. 1).

Some of it avoids immediately becoming landfill and reaches second-hand clothing markets. If this second-hand clothing is not sold in the United States market it is sent overseas to be sold in foreign second-hand markets (with the US exporting around 500,000 tons [450,000 tonnes] annually). Clothing that is not sold in foreign markets "becomes solid waste, clogging rivers, greenways, and parks, and creating the potential for additional environmental health hazards in [countries] lacking robust municipal waste systems" (Bick et al. 2018, p. 2).

The deep intertwining of identity and consumption for most people in western societies and increasing numbers of those in non-western societies means that restrictions on consumption are a direct threat to people's sense of self. They simply do not know how to be otherwise and their commitment to buying causes them to use it to address unhappiness through making more purchases. But the problems are not just a result of the power of the relationship between identity and consumption; insatiable consumption is intensifying due to the effects of fashion and, in particular, fast fashion. Accelerating consumption plays an important role in catastrophic human-caused environment change and there is no reason to believe that people will demand or even support political action to slow the rate of consumption to a truly sustainable level.

## Because They Have Been Brought Up to Believe in the End of the World, So Catastrophic Human-Caused Environment Change Is Not A Problem for Those in Positions of Political Power (Apocalypticism)

Another reason to believe that nothing meaningful will be done to address catastrophic human-caused environment change is that religious adherents who have been raised to expect the end of the world do not believe that anything can be done to prevent the 'end time'. Some actively desire it. With a relatively small number of Christians believing that efforts to prevent the end time are those of the antichrist or those who act for the antichrist. For them, political action to address the human causes of catastrophic environment change is either doomed to fail or wrong. If they do nothing else, the movements discussed in this section normalize the idea of the end of the world in the minds of their followers. While, as Torres wrote, "apocalyptic movements can be found on nearly every continent", he also observed that "convictions that the end is nigh are widespread in both the Christian and Islamic communities" (2016, p. 56).

This didn't mean that Christians and Muslims wanted the end to come. But it did mean that most of the people in these communities "'know' that life on Earth has a preordained expiration date, no matter what". They do not support efforts to address catastrophic human-caused environment change because "end-times believers would rationally perceive such efforts to be ultimately futile, and hence ill-advised (Barker and Bearce 2013, p. 269). One of the problems with efforts to address catastrophic human-caused environment change, for those committed to this view, was that it denied their Truth that, as Carr and colleagues put it, "God was in control and humans cannot do much about it." With several participants in their study dismissing "what they perceived as scientists and politicians claiming that humans could control the climate" (Carr et al. 2012, p. 288). They knew "that humans do not have the power to destroy the earth, so human-induced climate change would not be a cataclysmic event" (Carr et al. 2012, p. 293).

That God is in control provides another reason to accept the onset of the final hundred years of humanity. For those who believe in end times are predisposed to "ignore phenomena like global warming because they fail to fit the prewritten narratives of prophetic scripture" (Torres 2016, pp. 56–7). Examples of the last of these are those Christians who reject projections with respect to global warming because they predict a rise in sea levels. The profoundly religious residents of a Scottish fishing village Webster studied knew that "rising sea levels were an impossibility because God had covenanted not to allow it" (Webster 2013, p. 74).

Some, though, see catastrophic human-caused environment change as proof of prophecies concerning the end of times. As Torres suggested "many religions identify natural disasters as precursors of the Apocalypse. Famines, earthquakes, infestations, biodiversity loss, heat waves, landslides, and burning rain are all seen by certain traditions as harbingers of the end" (Torres 2016, p. 59). Indeed, "premillennial dispensationalists" believe "that earthly conditions must precipitously deteriorate prior to the Second Coming (marked by wars, famine, and environmental calamity)" (Barker and Bearce 2013, p. 268). Whereas other Christians quote 2 Peter 3:10: "the earth also and the works that are therein shall be burned up."

Teachings concerning the end time will seem increasingly relevant to many of those who witness the intensification of the effects of catastrophic human-caused environment change. So, as Torres suggested, catastrophic human-caused environment change "could actually reinforce religious people's... convictions that the world's end is imminent." This meant that global warming will not only "cement people's prior beliefs, it may well even lead to an increase in the number of religious adherents. After all, people often convert during periods of personal or societal stress, and history provides numerous examples of natural disasters boosting the membership of religions" (Torres 2016, p. 59).

Whereas some of those who believe in the end time do not desire or participate in action to address catastrophic human-caused environment change, others "see themselves as active participants in an End-Times narrative that's unfolding in real time... [and] believe it's their duty to either bring about Armageddon or foment the conditions necessary for the Apocalypse to take place" (Torres 2016, p. 56–7). Rather than simply resist attempts to reduce the human causes of catastrophic environment change, these believers seek to increase these causes to bring about the end time as quickly as possible. They "take it upon themselves to catalyze this process in an effort to hasten the moment of cosmic justice. ISIS, for example, is taunting the West into meeting it on the fields around Dabiq so that Armageddon can be fought, thereby knocking over the first domino of their Apocalypse" (Torres 2016, p. 58). At a more general level, according to Celso, the very "rise of the IS and its construction of the caliphate is presented… as a transcendental step marking the 'end times' and the inevitable destruction of the apostate and crusader forces" (2014, p. 1–2).

Christians and Muslims who believe in the end time are particularly important for this discussion because, according to a 2010 poll by researchers at the Pew Research Centre, Christianity and Islam were the fastest growing religions. Projecting from the findings of this study, Torres wrote "that there will literally be billions of believers in 2050 who will expect Armageddon to be just around the corner. … we can expect some 1.2 billion Christians with apocalyptic beliefs by the middle of this century. Another poll finds that apocalypticism is similarly widespread within the Muslim community" (Torres 2016, p. 59). Most Christians and Muslims will not take on an apocalyptic view, but "the total number of believers at the fringe is likely to increase as the total number of religious adherents grows. In a world with 9.3 billion people by 2050, approximately 8.1 billion of whom will be religious, 'the fringe' will constitute a rather sizable community" (Torres 2016, p. 60).

This effect may not be great. It represents another ideological effect, however, that contributes to political inaction with respect to addressing the human causes of catastrophic environment change. And the effect will increase if the people who believe in end times are in positions of political power or have a close relationship with those in positions of political power. And it will increase as more people come to accept apocalyptic faiths, of which Christianity and Islam will be the most significant. If nothing else, an important point to take away from this section is that we cannot assume that all people are seeking to avoid a trajectory that will take humanity to the beginning of its final hundred years and, instead, must accept that some are reconciled to this trajectory whereas others desire it.

## Because People in Countries of the Global South See Those in Countries of the Global North as Responsible for Catastrophic Human-Caused Environment Change (Postcolonialism)

Another explanation for the lack of a strong demand or even significant support for political action to address the human causes of catastrophic environment change is that people in countries of the Global South have legitimate grounds for believing that it is unfair to ask them to accept the costs associated with addressing those causes. This is important because, as Kartha and colleagues argued, "the historical inability to resolve issues of equity has arguably contributed to the slow rate of progress on climate action" (2018, p 118).

The inability to resolve issues of fairness or justice at the global level, as Kartha and co-authors suggested, means that "nations have been unable to address the question of how to equitably share the considerable effort that would match the climate change goals they have set, leading us to a regime based on nationally determined efforts that together fall short of those goals" (Kartha et al. 2018, p 118). Equity is just as much an issue within nations, though, and this has contributed to the larger equity problems associated with the impeding of international political action to address catastrophic human-caused environment change. As a failure to address equity issues within nations "has led important constituencies to see climate change mitigation—more so than climate change itself—as a threat to their livelihoods, their access to energy, and their freedoms." The result is "an audience readily disposed toward the anti-climate coalitions that have hindered ambitious action" (Kartha et al. 2018, p 118).

Having to deal with equity issues makes addressing the human causes of catastrophic environment change even more difficult as it moves beyond simply reducing or eliminating those causes as quickly as possible. The United Nations Framework Convention on Climate Change, for example, complicates and adds to the problem by calling "for a transition that 'contribute(s) to the goals of decent work for all, social inclusion and the eradication of poverty.' … And the Scottish Government is seeking a transition that 'promotes inclusive growth, cohesion and equality'" (Piggot et al. 2019, p. 1). Nothing here is intended to deny the validity of the equity issues that arise with respect to addressing the human causes of catastrophic environment change. The point is simply that the additional burdens associated with addressing the equity issues that arise in this context makes it very difficult, perhaps impossible, for those in governments to introduce and implement policies that will avert the path toward the final hundred years of humanity while also addressing those equity issues.

A significant problem arises at the global level, then, and in international forums in which those in countries of the Global South have been and will be asked to contribute to addressing the human causes of catastrophic environment change. Some commentators in countries of the Global South see forums in which representatives of countries of the Global South are put under pressure to participate in these international agreements as amounting to "environmental colonialism" (Agarwal and Narain 2012, p. 81). Others have simply argued "that climate change itself is unfair to developing countries" (Kong 2019, p. 105).

Resistance to being drawn into international agreements to address catastrophic human-caused environment change on the parts of those in countries of the Global South begins with the argument that greenhouse gases concentrations are an historical phenomenon to which they did not contribute. For Kong, "the rich life of people in developed countries is based on the high greenhouse gas emissions of their forefathers, so it is necessary for them to pay for it" (2019, p. 110). As Hurrell and Sengupta put it: "pointing to the consistent failure of the developed world to deliver either on their own emissions reduction commitments or on their promises on finance and technology, emerging powers have argued that the burden of solving climate change cannot now be passed on to them" (2012 p. 470).

But not only is catastrophic human-caused environment change a result of past actions of people in countries of the Global North, having to limit their growth to address it represents a present injustice for those in countries of the Global South. For "the bottom line is that climate change cannot be resolved by preventing or slowing the development of developing countries," according to Kong, "and the development right of developing countries cannot be undermined in the name of addressing global climate change" (2019, p. 106).

Another problem with calling for or forcing their cooperation is that those in countries of the Global South do not have the resources that countries of the Global North have to deal with catastrophic human-caused environment change. So not only do they have an historical responsibility to bear all, or most of, the costs associated with addressing the human causes of catastrophic environment change, people in countries of the Global North "also have sufficient funds and advanced technologies to cope with climate change" (Kong 2019, p. 110).

Further, "the relative backwardness of strength and technology, coupled with problems such as shortage of development capital… [and] heavy debt" mean that, when it comes to political action to address the human causes of catastrophic environment change, "developing countries have to pay the same amount of effort as countries of the Global North, or even more, but lose at the starting line. This is bound to widen, not narrow, the development gap between the North and the South" (Kong 2019, p. 111). The result, from the perspective of those in developing countries, is that "it is obviously unfair for those who are still at the stage of solving the problem of food and clothing to bear extra responsibilities, especially to pay for the consequences of past actions of those who have already led a higher quality of life" (Kong 2019, p. 111).

Yet another equity question arises with respect to accounting for the contributions of those in the Global South and North make to the human causes of catastrophic environment change. The calculation of national contributions and responsibilities for greenhouse gas emissions is a particularly contentious matter. Agarwal and Narain's response to a 1990 report by the World Research Institute (WIR), a Washington-based private research group, provides just one early example of the objections that those in countries of the Global South raise in this context.

In the first place, Agarwal and Narain rejected the WIR report because it made "no distinction between those countries which have eaten up this ecological capacity [to absorb CO2 and methane] by exceeding the world's absorptive capacity and those countries which have emitted gases well within the world's cleansing capacity" (2012, p. 84). Those in countries of the Global North have contributed to the causes of catastrophic environment change for decades longer than people in countries of the Global South have contributed to them. But this is not considered when those in countries of the Global South are being called upon to contribute.

The next problem, for Agarwal and Narain, is that the WRI report treated current emissions as the same even though they are related to very different activities. This brought them to ask whether we can "really equate the carbon dioxide contributions of gas guzzling automobile in Europe and North America… with the methane emissions of draught cattle and rice fields of subsistence farmers in West Bengal or Thailand?" Agarwal and Narain complaint, in this context, was that "no effort has been made in the WRI's report to separate out the 'survival emissions' of the poor, from the 'luxury emissions' of the rich" (Agarwal and Narain 2012, p. 82).

Further, according to Agarwal and Narain carbon sinks in India were absorbing more emissions than are being produced in India, which meant that India was not exceeding acceptable levels of greenhouse gas emissions. India's population in

1990 was "16.2 per cent of the world's population in 1990 [whereas] India's total production of carbon dioxide and methane amounted to only six per cent and 14.4 per cent, respectively, of the amount that is absorbed by the earth's ecological systems." In Agarwal and Narain's view, "India can double its total carbon dioxide emissions without threatening the world's climate. And if it controls its deforestation, then it can increase its carbon dioxide emissions from fossil fuels several times" (Agarwal and Narain 2012, p. 84). And such an increase would be necessary to resolve the gap between India and countries of the Global North.

A final problem with international agreements that bind those in countries of the Global South to adopt policies to address catastrophic human-caused environment change is the power of those who represent countries of the Global North in the formulation of those agreements. Kong argued that "developing countries are far behind the developed countries in comprehensive national strength, economy, science and technology level … and are bound to be in an obvious disadvantage status in the climate change governance game" (Kong 2019, p. 110). This left and leaves those in developing countries vulnerable to the "environmental colonialism" to which Agarwal and Narain referred.

As Hurrell and Sengupta suggested, "some of the most serious moral problems have to do not directly with the distribution of costs and benefits but rather with the absence of the political or institutional conditions for fair bargaining over climate change" (2012 p. 482). From their perspective, the institutions that have developed gave those who represent countries of the Global North an unfair advantage when it came to bargaining with those representing countries of the Global South. This "has reinforced unequal bargaining power between rich and poor and represents a threat of compound injustice" (Hurrell and Sengupta 2012 p. 482). As Hurrell and Sengupta argued, "the legitimacy of international institutions will remain seriously weakened to the degree that interstate inequalities continue to generate asymmetrical bargaining and continue to involve the domination of weaker parties by stronger" (2012 p. 482–3).

While most doubt in this context is cast upon international political forums, some of those in countries of the Global South are even drawn to question the science that provides the basis for discussions in those forums. For Agarwal and Narain the problems with the science on which the WIR report was based "emphasizes the fact that Third World nations must undertake their own research in this crucial area. They cannot depend on Western institutions to present a true picture of the global situation and safeguard our interests" (Agarwal and Narain 2012, p. 86). Though these (and others) are likely to be related problems.

The facts that those in countries of the Global North have a far greater historical and contemporary responsibility with respect to the problem make international agreements to truly address the human causes of catastrophic environment change unlikely. Those in countries of the Global South have and will raise several legitimate concerns with respect to equity that mean that they question the agreements and the targets they are given under these agreements. Equity concerns on the parts of those in countries of the Global South have meant and mean that the prospects for effective international cooperation are limited, especially in the years before the problems become so pronounced that an awareness that the final years of humanity have begun develops and desperation sets in. And by then it is probably going to be too late.

## Conclusion

Those in positions of political power are likely to be the targets of blame for inaction with respect to addressing catastrophic human-caused environment change. Even though they deserve much of the blame, and scientists some of the blame, they do not deserve all the blame. For they have not been subjected to sufficient popular pressure to introduce the fundamental changes necessary to address catastrophic human-caused environment change and avert a trajectory toward the final hundred years of humanity. Ideas concerning the normal, the natural, the desirable and the possible in the minds of people in western countries and, in some cases, non-western countries function to inhibit, where they do eliminate, a desire for the radical changes necessary to address the human causes of catastrophic environment change.

None of these ideological factors is enough, in itself, to produce inaction on the parts of those in government with respect to addressing the human causes of climate change. Together, however, they produce a situation in which recognition of a problem does not cause action to address that problem. Some of the factors, such as those associated with resistance to change, contemplating death in a society that is working to hide it, conceptualizing the problem and imagining solutions, affect many, if not most, people. Others, such as selfish individualism and consumerism, are more specific to those raised in western or capitalist societies. The spread of western values and capitalism, though, mean that selfish individualism and consumerism are becoming or defining the norms of an ever-growing number of people.

Fewer people expect, and in some cases desire, catastrophe. But it is important to recognize that there are some for whom catastrophic human-caused environment change is not the problem that it is for others. There are significant enough numbers of them to play a role in enabling ongoing political inaction with respect to addressing the human causes of catastrophic environment change, and their numbers will grow as the environment becomes increasingly unable to support the preservation of humanity. There are far greater numbers of those in countries of the Global South who see catastrophic human-caused environment change as the responsibility of those in countries of the Global North, however. For them it is up to those in countries of the Global North to do far more to avert a trajectory toward the final hundred years of humanity while allowing those in countries of the Global South to continue to pursue development.

## References

Agarwal, A., & Narain, S. (2012). 5 Global warming in an unequal world. In *Handbook of Climate Change and India: Development, Politics and Governance*, 81–8. https://doi.org/10.4324/9780203153284.

Ariès, P., *The hour of our death*, trans. Helen Weaver, London: Allen Lane, 1981.

Aronoff, K. (2019, 10 April) Fiscally-conservative Democrats have declared war on the Green New Deal, *The Guardian*. https://www.theguardian.com/commentisfree/2019/apr/10/fiscally-conservative-democrats-green-new-deal.

Barker, D. C., & Bearce, D. H. (2013). End-times theology, the shadow of the future, and public resistance to addressing global climate change. *Political Research Quarterly*, *66*(2), 267–279. https://doi.org/10.1177/1065912912442243.

Bartlett, B. (1994). The high cost of turning green. *The Wall Street Journal*, *14*(0).

Baumeister, R. F., Bratslavsky, E., & Muraven, M. (2018). Ego depletion: Is the active self a limited resource?. In *Self-Regulation and Self-Control* (pp. 24–52). Routledge.

Bick, R., Halsey, E., & Ekenga, C. C. (2018). The global environmental injustice of fast fashion. *Environmental Health*, *17*(1), 92. https://doi.org/10.1186/s12940-018-0433-7.

Brandt, M. J., Sibley, C. G., & Osborne, D. (2019). What Is Central to Political Belief System Networks?. *Personality and Social Psychology Bulletin*, 0146167218824354. https://doi.org/10.1177/0146167218824354.

Brick, C., & van der Linden, S. (2018). Yawning at the apocalypse. *The Psychologist*, 30–35. https://thepsychologist.bps.org.uk/volume-31/september-2018/yawning-apocalypse.

Carr, W., Patterson, M., Yung, L., & Spencer, D. (2012). The Faithful Skeptics: Evangelical Religious Beliefs and Perceptions of Climate Change. *Journal for the Study of Religion, Nature & Culture*, *6*(3), 276–299.

Celso, A. N. (2014). Dabiq: IS's apocalyptic 21st century jihadist manifesto. *Journal of Political Sciences & Public Affairs*, *2*(4), 1–4. https://doi.org/10.4172/2332-0761.1000e111.

Dang, J., Zerhouni, O., Imhoff, R., Jia, L., Giacomantonio, M., Sevincer, A. T., … Schiöth, H. B. (2019, January 14). Multi-Lab Replication Reveals A Small but Significant Ego Depletion Effect. *PsyArXiv*. https://doi.org/10.31234/osf.io/cjgru.

Elliott, A. (2006). A girl's guide to retail therapy. New York: Barnes and Noble Publishing.

Fahrenthold, D. A. (2009, Dec 14). We're not so good at long-term, global problems. *The News Journal* Retrieved from http://libproxy.murdoch.edu.au/login?url=https://search-proquest-com.libproxy.murdoch.edu.au/docview/275286827?accountid=12629

Fiske, A. P. (1991. The cultural relativity of selfish individualism: Anthropological evidence that humans are inherently sociable. In M. Clark (Ed.), *Review of Personality and Social Psychology 12: Altruism and Prosocial Behavior*. London: Sage.

Friend, C. (2004). Social contract theory. Internet Encyclopedia of Philosophy. https://www.iep.utm.edu/soc-cont/

Gerhardt, S. (2010). *The selfish society: How we all forgot to love one another and made money instead*. London: Simon & Schuster.

Good, P. (2018). Evidence the US autism epidemic initiated by acetaminophen (Tylenol) is aggravated by oral antibiotic amoxicillin/clavulanate (Augmentin) and now exponentially by herbicide glyphosate (Roundup). *Clinical nutrition ESPEN*, *23*, 171–183

Hirschauer, N., Lehberger, M., & Musshoff, O. (2015). Happiness and utility in economic thought—Or: What can we learn from happiness research for public policy analysis and public policy making?. Social Indicators Research, 121(3), 647–674.

Hornsey, M. J., Harris, E. A., & Fielding, K. S. (2018). Relationships among conspiratorial beliefs, conservatism and climate scepticism across nations. *Nature Climate Change*, *8*(7), 614.

Hurrell, A., & Sengupta, S. (2012). Emerging powers, North–South relations and global climate politics. *International Affairs*, *88*(3), 463–484.

Irwin, C. (2018). Emotional Outlet Malls: Exploring Retail Therapy. *BU Well*, *3*(1), 8.

Jacques, M. (2002). The age of selfishness. *The Guardian*, *5*. https://www.theguardian.com/comment/story/0,3604,805161,00.html

Jost, J. T. (2015). Resistance to change: A social psychological perspective. *Social Research: An International Quarterly*, *82*(3), 607–636.

Kartha, S., Caney, S., Dubash, N. K., & Muttitt, G. (2018). Whose carbon is burnable? Equity considerations in the allocation of a "right to extract". *Climatic Change*, *150*(1–2), 117–129.

Kazim, A. (2018). The Emergence of Hyper-Consumerism in UAE Society: A Socio-Cultural Perspective. *Perspectives on Global Development and Technology*, *17*(4), 353–372.

Kinder, D. R., & Kalmoe, N. P. (2017). *Neither liberal nor conservative: Ideological innocence in the American public*. University of Chicago Press.

Kong, Q. (2019). Justice of Climate Change Governance in the Changing World--Discussion Based on the North-South Relations. *Ekoloji*, *28*(107), 105–113.
Lacroix, K., & Gifford, R. (2018). Psychological barriers to energy conservation behavior: The role of worldviews and climate change risk perception. *Environment and Behavior*, *50*(7), 749–780.
Lamont, M., Park, B. Y., & Ayala-Hurtado, E. (2017). Trump's electoral speeches and his appeal to the American white working class. *The British journal of sociology*, *68*, S153-S180.
Lasch, C. (2018). *The culture of narcissism: American life in an age of diminishing expectations*. WW Norton & Company.
Latour, B. (2014). War and peace in an age of ecological conflicts. *Revue juridique de lenvironnement*, *39*(1), 51–63.
Macpherson, C. B., & Cunningham, F. (1962). *The political theory of possessive individualism: Hobbes to Locke* (p. 331145263). Oxford: Clarendon Press.
Maniates, M. (2002) 'Individualization: plant a tree, buy a bike, save the world?', in Confronting Consumption, eds. T. Princen, M. Maniates & K. Conca, Cambridge: Massachusetts Institute of Technology.
Mason, L. (2018). Ideologues without issues: The polarizing consequences of ideological identities. *Public Opinion Quarterly*, *82*(S1), 280–301.
Merkley, E., & Stecula, D. A. (2018). Party elites or manufactured doubt? The informational context of climate change polarization. *Science Communication*, *40*(2), 258–274. https://doi.org/10.1177/1075547018760334.
Panno, A., Carrus, G., & Leone, L. (2019). Attitudes towards Trump Policies and Climate Change: The Key Roles of Aversion to Wealth Redistribution and Political Interest. *Journal of Social Issues*, 75(1), 153–168. https://doi.org/10.1111/josi.12318.
Piggot, G., Boyland, M., Down, A., & Torre, A. R. (2019). Realizing a just and equitable transition away from fossil fuels. *Development, 2016*, 202033. http://www.indiaenvironmentportal.org.in/files/file/realizing-a-just-and-equitable-transition-away-from-fossil-fuels.pdf. Accessed 7 Jan 2020.
Rick, S. I., Pereira, B., & Burson, K. A. (2014). The benefits of retail therapy: Making purchase decisions reduces residual sadness. *Journal of Consumer Psychology*, *24*(3), 373–380.
Sampson, E. E. (2000). Reinterpreting individualism and collectivism: Their religious roots and monologic versus dialogic person–other relationship. *American Psychologist*, *55*(12), 1425.
Sklair, L. (2012). Culture-Ideology of Consumerism. *The Wiley-Blackwell Encyclopedia of Globalization*.
Sonne, J. W. H., & Gash, D. M. (2018). Psychopathy to Altruism: Neurobiology of the Selfish-Selfless Spectrum. *Frontiers in psychology*, *9*, 575.
Thompson, S. (2012). The micro-ethics of everyday life: ethics, ideology and anti-consumerism. *Cultural studies*, *26*(6), 895–921.
Torres, P. (2016). Apocalypse soon? How emerging technologies, population growth, and global warming will fuel apocalyptic terrorism in the future. *Skeptic (Altadena, CA)*, *21*(2), 56–63.
Varcin, K. J., Alvares, G. A., Uljarević, M., & Whitehouse, A. J. (2017). Prenatal maternal stress events and phenotypic outcomes in Autism Spectrum Disorder. *Autism Research*, *10*(11), 1866–1877.
Wagoner, B., Moghaddam, F. M., & Valsiner, J. (Eds.). (2018). *The Psychology of Radical Social Change: From Rage to Revolution*. Cambridge University Press.
Webster, J. (2013). The eschatology of global warming in a Scottish fishing village. *The Cambridge Journal of Anthropology*, *31*(1), 68–84. https://doi.org/10.3167/ca.2013.310106.
Xavier, M. (2018). *Subjectivity, the Unconscious and Consumerism*. Springer.

# Chapter 4
# The Climate Change Denial Industry

What has been labeled 'the climate change denial industry' has played a significant role in preventing or slowing action to address the human causes of catastrophic environment change. So, as was noted in Chap. 3, this 'industry' is one of the factors that have contributed to the maintenance of a trajectory toward humanity's final hundred years. The effectiveness of the industry is also a reflection of the vulnerability to challenge that scientists create when they represent science as a source of absolute Truth. But the effectiveness of the industry also illustrates the power of specific industries, organizations and individuals to influence public perceptions and key actors in positions of political power. And, finally, the changes in strategy on the parts of those in this industry, as the signs of catastrophic human-caused environment change become increasingly obvious, provides yet another element of the politics of the final hundred years of humanity.

As Dunlap and Jacques wrote, "no sooner had human-caused global warming been placed on the public agenda, … than an organized campaign to deny its reality and significance was launched" (Dunlap and Jacques 2013, p. 700). The main strategies that have been adopted by those in the climate change denial industry are to claim that catastrophic climate change is not happening, to manufacture uncertainty as to the role that humans have played in causing catastrophic environment change – in part by suggesting that 'all the evidence is not in' with respect to catastrophic human-caused environment change, and to argue that regulating the industries that are responsible for climate change comes at too great a cost. In pursuit of the first two strategies, the climate change denial industry mounted "a full-scale assault on… climate science, the IPCC, scientific organizations endorsing [human-caused global warming], and even individual scientists" (Dunlap and Jacques 2013, p. 700).

As one of the authors of the article that introduced the 'hockey stick' graph, which illustrating the increasing global average surface temperature (Mann et al. 1998, p. 779), Michael Mann found himself the target of those in the climate change denial industry. Their goal was "to discredit the graph—and discredit him—as a means of dismissing the case for human-caused climate change altogether." This strategy reflected "a cynical belief that if the science behind the hockey stick curve



I. Cook, *The Politics of the Final Hundred Years of Humanity (2030-2130)*,
https://doi.org/10.1007/978-981-15-1259-9_4

is dismissed, the whole climate change debate will collapse like a house of cards" (Mann and Olson 2018, p. 48).

Another strategy used by the climate change denial industry includes "fake grass roots campaigns that are coordinated by fossil fuel interest groups (and the politicians who support them) to inject doubt into the public discourse." As part of these campaigns, front groups are created, "some of which employ actors to portray defiant citizens, others of which recruit apostolic scientists to publicly undermine their peers—to openly contest the science of climate change, in turn creating the illusion of a groundswell of opposition" (Mann and Olson 2018, p. 49).

Technically, this industry should be referred to as the 'climate change skepticism industry' and not the 'climate change denial industry'. This is its most common designation, however, though the more important reason to refer to it as the climate change denial industry is because those involved in this industry are inconsistent in their skepticism. In short, the industry's members are not skeptical about their own position. Many are happy to describe climate scientists who affirm catastrophic human-caused environment change as doing so based on false or incomplete data, biased data selection or other processes that reproduce the (False) mainstream consensus. But those in the climate change denial industry present their position as the Truth and do not consider the processes that support the reproduction of their Truth.

Thus, Emeritus Professor of Chemistry and Science Studies at Virginia Polytechnic Institute and State University (Virginia Tech) and self-described 'climate change skeptic', Henry Bauer, is right to suggest that "scientists need to earn a living, and that can mean a conflict of interest between doing what an employer or patron wants as against doing the best science" (Bauer 2017, p. 9). Bauer's problem, however, is that he does not extend this observation to scientists working in the climate change denial industry. They can be no more certain that they have absolute Truth than any other scientist. Bauer is also right to argue that "institutions of science… are no more willing to brook criticism of their beliefs and dogmas than other authoritative social institutions are or ever have been" (Bauer 2017, p. 13). He is simply wrong to think that this applies only to mainstream climate science and not to scientists whose wages and financial support come from those who seek to promote climate change skepticism and who are bound by their own institutions' pressure to resist criticism.

Grasso went further, though, to argue that those who deny climate change have become hostile toward mainstream climate scientists. In his view, "a political context able to politicise the scientific orthodoxy, not simply to dispute it, was carefully orchestrated with the money contributed by some IOCs [international oil companies]" (Grasso 2019, p. 113). In what Grasso saw as a new form of denial that shifted from dispute to attack, those who deny climate change "have been able to portray climate change as an issue created by climate scientists for climate scientists, as a fabrication to keep alive an obscure, yet sizeable, technoscientific elite, as well as for the good of the notorious pro-big government and higher tax environmentalists" (Grasso 2019, p. 113). The result is to foster "a most intolerant scepticism among denialists… doubts about the evidence of climate science have been skilfully replaced by doubts about the motives of the people who study climate change and communicate scientific findings" (Grasso 2019, p. 113).

And if it is not the motives of climate scientists that are being questioned, it is the motivations of those in positions of political power that come to be questioned. Even though they are often more mouthpieces for, rather than the sources of, climate change denial, leaders of populist parties often treat policies to address catastrophic human-caused environment change as part the elite's conspiracy against the people. As Schaller and Carius wrote, "overtly denialist parties portray their people as victims of secret plans by leading political actors… Not uncommonly used is the claim that anthropogenic climate change is an invented theory used to draw (financial) resources out of the public…" (2019). These parties combine "nationalism and authoritarianism… with anti-elitism to construct a world view in which 'the people' are ruled by a corrupt and illegitimate liberal, cosmopolitan elite. While the main targets here are immigration…, the climate change agenda fits in well as collateral damage" (Lockwood 2018, p. 726).

In addition, and again technically, this industry should also be called the 'human-caused climate change denial industry' because its members need only contend that environment change is not caused by human activity. As, Oreskes and Conway noted, the initial strategy adopted by those in the industry "wasn't to deny the fact of global warming, but to blame it on the Sun" (2011, p. 186). Many of those who fund climate change denial organizations and individuals are involved in industries that they are seeking to protect from regulation, or even termination, because they contribute to the human causes of catastrophic environment change. Their goal is simply to defend their industries and they require only that the principal human cause of catastrophic environment change, global warming, is not taken to result from human activity. Those in the climate change denial industry have often, if not usually, gone further to deny that global warming is occurring. But doing so is not essential to protecting those industries that contribute to global warming, which is important to maintaining a trajectory toward the final hundred years of humanity.

The final preliminary question that arises in this context concerns whether referring to those who refuse mainstream as participating in an 'industry' is more than simply a rhetorical device. Using the term 'industry' might be considered hyperbolic, but a high level of coordination has been found to exist amongst many of those involved in denying climate change. Carroll and co-authors identified the operation of a "denial regime [that] along with political allies… promote and implement convivial policies, and an ideological apparatus of think tanks and policy groups…" (Carroll et al. 2018, p. 444).

They believed that their study's "findings show a pervasive pattern of carbon-sector corporate influence, spanning across domains of civil society to form a single, connected network" (Carroll et al. 2018, p. 444). They found that groups representing business organizations in various industries responsible for the causes of catastrophic environment change interlocked to "create pathways of influence across sectors." In their view, "business councils … stand out as places where business leaders, often active in other CSOs [civil society organizations], collaborate in crafting a shared agenda for big business. Their extensive affiliations convey the corporate world view to policy-planning groups, universities, and research institutes…" (Carroll et al. 2018, p. 445). Hornsey, Harris and Fielding's depiction is much like Carroll and co-authors'. To them, "the case has been made that a network of vested interests, think tanks and ideologically driven scientists has succeeded in permeating the conservative political culture, locating climate science within the same culture wars that have traditionally divided people on issues such as gun control and abortion" (Hornsey et al. 2018, p. 616).

Michaels saw the climate change denial industry in a much broader way than many other analysts have presented it. He suggested that it is not so much a distinct climate change denial industry but a continuation of an industry engaged in preventing government and consumer action to limit or eliminate industries that harm human health. In his view, "the pseudo-scientific and political opposition to acknowledging climate change is aligned—ideologically, tactically, and now politically—with the same gang that has specialized in manufacturing scientific uncertainty since the 1950s. It is the same crowd, funded by the same money, employing the same tactics" (Michaels 2019, p. 7).

Those who participate in this 'crowd' had already "perfected the art of manufacturing uncertainty about the proven dangers of exposure to things like cigarette smoke, lead paint, industrial chemicals, identical strategies". This is now being directed to "denying and undermining the science that documents the atmospheric buildup of greenhouse gases, and the impacts of this accumulation on life on this planet" (Michaels 2019, p. 7). In Michaels' account, then, it is not so much a climate change denial industry as it is a science denying industry that turned its hand to climate change denial. Oreskes and Conway support Michaels' characterisation of the climate change denial industry (2011).

This chapter is in four parts. It begins by accounting for the level of consensus that exists amongst climate, and other, scientists concerning the existence and significance of catastrophic human-caused environment change. Next it provides evidence about the extent of public skepticism toward the existence or significance of human-caused environment change. Then it moves to explaining why, even though scientists are convinced that catastrophic human-caused environment change is real and imminent and urgent action is necessary to avoid disaster, so many people are not convinced of these things, especially the need for urgent action.

The answer presented here is that, in many cases, this is a result of the power of the climate change denial industry, which makes some understanding of the way that this industry exercises its power important to an account of the politics of the final hundred years of humanity. Finally, in keeping with the occasionally speculative character of this part of the book, some observations are offered concerning what those involved in the climate change denial industry are likely to do during the final hundred years of humanity.

If the science on which this book relies is accurate, the catastrophic effects of human-caused environment change will become more obvious as the environment moves toward and into a state in which humanity will no longer be possible. This will mean that those involved in the climate change denial industry will require new strategies. Two strategies seem most obviously applicable under these conditions.

One is to refuse, or dispute, the 'human-caused' aspect of catastrophic human-caused environment change. Denying that catastrophic environment change has begun will be rendered impossible. But denying that it is human-caused remains available. An obvious strategy for those in the climate change denial industry, then, is to acknowledge environment change, even catastrophic environment change, but to deny that it, or much of it, is caused by human activity.

Despite growing evidence to the contrary, the view that environment change is not a result of human activity persists. As recently as 2017, Mann and Olson felt compelled to address this view. Their article "demonstrated that the years 2014, 2015, and 2016 were the warmest years on record (in ascending order) and estimated that it is highly unlikely that this trend was not due to human behavior" (2018, p. 48). But this does not mean that the claim that catastrophic environment change is not human caused has disappeared. Myriad websites continue to publish lists of reasons why climate change is not caused by human activity.

But even if members of the climate change denial industry are forced to accept that climate change is human-caused they can focus on the costs involved in doing something about it to argue for inaction. Thus far, this strategy has not been a prominent aspect of the position put by those in the climate change denial industry. The problem with this strategy, for them, is that it requires admissions both that catastrophic environment change is occurring and that it is human-caused. This negates previous strategies of denying that catastrophic environment change is occurring and of denying that it is human-caused. This makes adopting the position that nothing should be done because it costs too much to do anything meaningful a fundamental shift in approach. But this strategy will become necessary for stopping governments from regulating, or even closing, the industries that contribute most to the human causes of catastrophic environment change.

## Part One: Ignoring Science

### *Climate Scientists Are Not Skeptical About World Threatening Human-Caused Environment Change*

Suggesting that climate scientists disagree about whether catastrophic environment change is occurring and is human caused is false. This is not to claim that they have an absolute Truth, for, as discussed in Chap. 2, there may be an element of groupthink at work here. The group is sufficiently large, though, and the fabrication of evidence sufficiently difficult that it would be unimaginably arrogant to simply reject their consensus. Just as it would be profoundly unwise to ignore this consensus, given the consequences that flow from being wrong about climate scientists being right. The matter at hand is, after all, the end of humanity.

A necessary starting-point for an analysis of the climate change denial industry is recognition of the fact that climate scientists are close to unanimous in believing that catastrophic *human-caused* environment change is occurring and that it represents a profound and imminent threat. In short, whether they are right or wrong, practically no climate scientist has significant doubt on this matter. Yet doubt exists in the wider community and, further, some deny that catastrophic human-caused environment change is occurring and is a profound and imminent threat. This doubt must be produced, some suggest 'manufactured', by those outside the scientific community and conveyed to voters who support politicians and parties that deny that catastrophic human-caused environment change is occurring and is a profound and imminent threat.

Much of the research done on the consensus amongst scientists about human-caused environment change has been about climate change, and often the focus is on global warming as a synonym for climate change. So, the focus of this chapter will be on climate change, and occasionally global warming, even though it is important to remember that species depletion and extinction and increasing environmental toxicity means that it is the environment, and not just the climate, that is changing. And when it comes to climate change, climate scientists believe that it is occurring and is worsening; and they do so with a great deal of confidence.

According to the authors of a study released in 2013, and there is no reason to think that anything has changed, over 97% of the papers published in authoritative climate science journals supported the position that humans were causing global warming. As the authors of this study put it "the number of papers rejecting the consensus… is a vanishingly small proportion of the published research" (Cook et al. 2013, p. 1). A later study of a sample of the less than 3% of papers whose authors did not accept human-caused global warming found a pattern of flaws in the methods used in these articles and other mistakes that rendered their findings invalid (Benestad et al. 2016, p. 699).

An analysis of surveys of climate scientists who were publishing in the field found that either nine out of ten or ten out of ten of these scientists were certain that human activity was causing global warming. The difference in the results from the studies related to the exact question that was asked and the sample of climate scientists being used. These findings were supported by several other studies (see Cook et al. 2016). So journal articles written by climate scientists and surveys of climate scientists' opinion on the matter show their almost unanimous agreement that global warming was happening and was caused by human activity.

And it is not just climate scientists who believe that human activity is causing global warming. Ninety-three percent of biophysical scientists working in 12 of the biggest universities in the United States said that they believed that the average global temperature had risen (when they were surveyed in 2014). Three and a half percent were not sure, and not being easily convinced is something we should expect and want from scientists (Carlton et al. 2015, p. 1). The study in question was of scientists whose research field was not climate science and the 3% who thought that the average global temperature had stayed the same or fallen disputed the validity of climate science itself. So, some form of rivalry amongst scientists may well have been involved here – especially in light of the public prominence being given to the work of climate scientists. Further, of the scientists in this survey who were convinced that the average global temperature had risen, only 1% of them were not sure that this was mainly the result of human activity (Carlton et al. 2015, p. 9).

## *But Enough People Are Skeptical About World Threatening Human-Caused Environment Change to Have Contributed to Inaction*

Despite the overwhelming consensus amongst climate scientists, though, there are still significant numbers of people who are either uncertain or who think that climate change is not real. Just as importantly, given that inaction with respect to addressing the human causes of catastrophic environment change will lead to the final hundred years of humanity, many more believe that preventing catastrophic human-caused environment change should not be the highest priority for those in government. These people play an important, if not central part, in the ongoing failure to address the human causes of catastrophic environment change.

Those who think that climate change is not happening and those who think that climate scientists do not agree about the reality of climate change are interesting for this chapter. And they are particularly interesting given the effect of information about the true level of the scientific consensus on the questions of the existence of and human responsibility for catastrophic environment change. For once they find out about this consensus, many skeptics change their position with respect to the need to address climate change. The fact that many change their view when provided with this information indicate the existence of circuits of information that flow independently of the evidence about the scientific consensus around catastrophic human-caused environment change.

Surveys of people's views on the reality of climate change have been run in a variety of countries – though the focus is often on those in western countries. The US and Australia, where over 20% of people have been found to think that climate change is not occurring, are outliers. The researchers at the Pew Research Centre put the number in the United States at 20% in 2016 (Funk and Kennedy 2016); whereas those at the Texas University Energy Poll found that 21% of Americans thought climate change was not real in 2016 and that the number who thought this increased to 25% in 2017 (UT at Austin 2017). A survey by Australia's CSIRO found that 22% of the Australians polled in 2014 thought that climate change was not real (Leviston et al. 2015). This can be compared with 10% of Canadians (The Environics Institute 2015), 6% of Germans, 2% of those in the UK, 1% in France and less than 1% in Norway (Steentjes et al. 2017, p. 19).

People who do not accept that climate change is real are interesting in this context because they cannot have come to this conclusion for themselves. They are not climate scientists, so they cannot have done the necessary research. If they had listened to climate scientists, then, due to the powerful consensus amongst climate scientists, they would have been compelled to agree that human-caused climate change was occurring and worsening. Even though some non-scientists might think that they rely upon *common* sense, they are in a significant minority. So even if it is sense, it is not common sense.

And even though they might have wanted to refuse the reality of climate change (in a way, most people do), this is not the same as refusing to accept that it is real. Somebody has, or some-bodies have, put this idea into people's minds. Those some-bodies are the subject of this part of the story of the final hundred years of humanity. These some-bodies are interesting both because of their power in creating doubt and even denial and for how they will adjust when the evidence with respect to catastrophic human-caused environment change cannot be denied.

Those who doubt or deny the existence of climate change are not the only indicators that the consensus amongst climate scientists that human-caused climate change is real and is worsening is not compelling people to seek and pursue collective action to address the human causes of catastrophic environment change. The numbers of people who are unaware of this consensus or who doubt that a consensus exists are also interesting. Even though the following statistics relate only to Americans and are likely only to otherwise apply to Australians (who, as noted above, also have higher numbers of climate change deniers), they tell a fascinating and important story. The first statistic is that a third of the American public thought that scientists disagree about the validity of global warming (Wiens 2016, p. 71). The second statistic is that in 2015 only 12% of the American public knew that between nine and ten out of ten climate scientists accept the reality of global warming (Cook et al. 2016, p. 6).

Once again, the question that arises here concerns how the participants in these studies could be so misguided with respect to the views of climate scientists. Dismissing them as ill-informed and ignorant is too easy and is unhelpful. Those involved in the studies are undoubtedly ill-informed with respect to the consensus amongst climate scientists. But few of them want to be wrong. They may want to deny the truth about climate scientists, as this is an unpleasant truth (and more than an inconvenient truth). For some the scientific consensus concerning catastrophic human-caused environment change does not feel right, and there is reason to believe that, as Carlton and colleagues put it, it is not just about "the 'facts' of climate change, but for the 'feel' of it, as well" (2015).

Ultimately, though, whether they want it or it feels right, studies indicate that many of those who doubt or deny climate change are not choosing to ignore the facts. They are often simply unaware of them. A study by Van der Linden and colleagues showed that doubt or disbelief with respect to catastrophic human-caused environment change is often a result of the right information not getting to the doubters and deniers. Their study makes it clear that people who thought that human-caused climate change was not real will revise their views when informed of the extent of the consensus amongst climate scientists that human-caused climate change was real (van der Linden et al. 2016, p. 15).

And, according to this study, it does not matter whether the people who once thought climate change was not real supported the Democrats or the Republicans. Of course, given that more supporters of the Republican Party think that climate change is not real than supporters of the Democratic Party, more of them were affected by being told about the consensus amongst climate scientists. But even Fox News watchers and people predisposed to conservative views changed their opinions when told of the scientific consensus (van der Linden et al. 2016, pp. 15–6).

## Part Two: The Climate Change Denial Industry

The study in which Americans changed their view that human-caused climate change was not real when told that nine to ten out of ten climate scientists believed that it was real is important for this discussion. It is somewhat more important than the fact that so many Americans were not aware of the consensus. And it is a great deal more important than the fact that one in five Americans think that human-caused climate change is not real. For it shows that the problem is one of a lack of information, rather than a desire to believe in false information (crucially, that climate scientists are not close to unanimous with respect to the existence of catastrophic human-caused environment change). But all of these findings are important because they point to something affecting the thinking of all those who believe that human-caused climate change is not real. This brings us to the ongoing efforts of those involved in the climate change denial industry.

The key actors in this industry, of course, are those who fund it. The industry could not exist without the billions of dollars that are provided to the miniscule number of scientists who deny the validity of mainstream climate science and its projections concerning catastrophic human-caused environment change. Under normal circumstances, i.e., those in which monied and powerful interests are not arranged to amplify the voices of climate scientists who dispute the mainstream, these scientists' voices would not be heard as widely as they are currently heard.

More importantly the voices of deniers would not drown out the voices of scientists seeking to inform people that scientists do not disagree concerning the validity of the science with respect to climate change and that, on the contrary, nine to ten out of ten of them believe that global warming is real. Almost all the academic research into the funding of the climate change denial industry has been focused on the United States of America, which is the literal, and not just spiritual, home of climate change denial. Hornsey, Harris and Fielding believe it possible to "argue that it is only in the United States that climate scepticism has insinuated itself so successfully within the elite of a mainstream political party, and that it has formed such a high-profile role within the political rhetoric" (2018, p. 616).

The representation of the climate change denial industry developed here, then, has a justifiably distinctive American bent. The industry exists in other countries, mostly western countries, but think tanks in the European Union do not have to disclose the sources of their funds, so conducting research into their financial support is difficult. And some of the money for European climate denial think tanks is alleged to come from US sources. For these reasons, then, the focus here will be on corporate funding of climate change denial in the US.

The foundations that have provided the bulk of the money were associated with Koch Industries and ExxonMobil. At least, we can be certain that they were between 2003 and 2007. After that these foundations ceased making publicly traceable contributions. Instead, they began giving money to foundations that then passed this money on to organizations involved in climate change denial. According to Farrell, the influence of "private individual and corporate donors from across the political spectrum… has grown in recent years with the rapid expansion of untraceable donor-directed philanthropy enabling actors to give anonymously via pass-through organizations such as DonorsTrust and Donors Capital Fund" (Farrell 2019, p. 1). One result is that while "more than $410 Billion dollars are donated each year in the US alone, only 5%… is from corporations" (Farrell 2019, p. 2).

These foundations were created to conceal the sources of their funds, of course. But there is a powerful correlation between the significant decline of traceable funding from foundations with an interest in promoting climate change denial and rapid growth of funding from sources such as DonorsTrust and Donors Capital Fund (Brulle 2014, p. 692). Even though this confluence occurs in a variety of contexts, "philanthropy and misinformation… have converged most clearly around the issue of climate change, where scholars have identified an empirical relationship between industry-led political philanthropy and the large-scale production and diffusion of scientific misinformation about climate change" (Farrell 2019, p. 1).

Most of the funds that are donated to the climate change denial industry go to supporting research organizations, or think tanks, that produce websites and reports that are made available online and are regularly taken up in media reporting – often governed by principles of 'balanced reporting' that, as will be discussed below, require authorizing the voices of those who deny climate change.

Most of these organizations are in the US, of course. A study published in 2014 identified 118 US organizations involved in climate change denial (Brulle 2014, p. 684). Only 91 of them, however, could be tracked through data from the Inland Revenue Service. Together they received over $7 billion in funding between 2003 and 2010, or more than $900 million a year (Brulle 2014, p. 685). Climate change denial was not the only focus of many of these organizations, so not all of their funding was devoted to climate change denial. But even a modest fraction of it remains a vast sum of money.

Think tanks funded to promote climate change denial are important because they take the lead in the way that climate change denial is presented. They do this by developing the arguments that others use to deny climate change, or human caused climate change. This is important because when arguments that have been developed in climate change denial think tanks are repeated by media personalities, politicians, bloggers and members of the public, they gain credibility (Farrell 2016, p. 97).

McCright and co-authors have suggested that the climate change denial think tanks played a crucial role in the changing the crucial question in the climate conversation. This was moving the question from "How should we deal with climate change?", which dominated the conversation during the mid-1990s, to "Is climate change a real problem?", which became the crucial question from the 2000s and continues to dominate the conversation (McCright et al. 2016, p. 186). This change in question was not just important for the voters who support political parties and media organizations that engage in denial. It also helps to sow a seed of doubt in the minds of those who think it may be real and affects their calculations with respect to the cost of the economic and social changes that would be necessary to address catastrophic human-caused environment change.

The arguments that have been developed in climate change denial think tanks are most likely to reach the public via a conduit of 'media personalities' and politicians. Brulle characterized the situation as akin to a stage production. Audiences see the performers. They do not see the investors and the others behind the scenes, however, who make the performance possible. So, "just as in a theatrical show, there are stars in the spotlight. In the drama of climate change, these are often prominent contrarian scientists or conservative politicians…" (Brulle 2014, p. 693). But these people "are only the most visible and transparent parts of a larger production. Supporting this effort are directors, scriptwriters, and, most importantly, a series of producers, in the form of conservative foundations" (Brulle 2014, p. 693).

As Mann and Olson pointed out, the climate change denial industry's strategies include creating "fake grass roots campaigns… to inject doubt into the public discourse." In many cases, the industry, with the support of compliant politicians, forms "front groups—some of which employ actors to portray defiant citizens, others of which recruit apostolic scientists to publicly undermine their peers—to openly contest the science of climate change, in turn creating the illusion of a groundswell of opposition" (Mann and Olson 2018, p. 49).

The efforts of those in climate change denial think tanks and others spreading the denial message have contributed to climate change denial becoming embedded in the ideology of many of the world's conservative parties. That acting to address catastrophic human-caused environment requires fundamental, at the root, change means that it is anathema to those involved in conservative parties. It has certainly become part of the agenda of most populist parties and some major conservative parties, with the US Republicans being the only major conservative party to almost explicitly require climate change denial on the parts of its representatives.

In general, though, the principal determinant of whether conservative parties will engage in climate change denial is their country's possession of significant quantities of fossils fuels. According to Båtstrand (who surveyed the manifestos of conservative parties in the UK, Norway, Sweden, Spain, US, Canada, New Zealand, Germany and Australia), "much of the literature… [anticipated] a conservative aversion against environmental measures as such. With regard to climate change, this is only the case with the U.S. Republican Party and hence not representative of conservative parties as a party family" (2015, p. 555).

Further, Båtstrand found no support for the assumption "that conservative parties influenced by neoliberalism have difficulties backing new eco-taxes and state regulations…" (2015, p. 555). Generalizing with respect to the positions and policies adopted by conservative parties he studied was difficult, in Båtstrand's view. One conclusion he drew from his study was that "the traditional left/right dimension has not absorbed climate change to the extent that conservative parties support only those climate measures based on free markets … conservative climate policies utilize policy parts from the left as well as the right on the state/market dimension" (Båtstrand 2015, p. 555). Båtstrand was willing, however, to entertain "the notion of a conservative pro-business position followed by natural resources playing a role in shaping climate policies. The conservative parties do not intend to challenge the fossil industry if the respective countries have vast reserves of fossils" (2015, pp. 555–6).

Populist parties tend to take up the ideas generated within the climate change denial industry more consistently than conservative parties, though, and provide a readier conduit for the transmission of the ideas generated within climate change denial think tanks. In their study of European right-wing populist parties, Schaller and Carius found that "seven of the 21… are sceptical of the scientific consensus on human-induced climate change or are overtly denying it. Another 11 parties are inconsistent in their communication or remain completely silent on the problem" (Schaller and Carius 2019, p. 10). The two right-wing populist parties that explicitly supported the consensus on climate change (Hungary's *Fidesz* and the Latvian *National Alliance*) were both parts of their respective national governments, which may have played a role in their different position on this issue.

Journalists have also played an important role in disseminating climate change denial, and not merely journalists working for conservative media organizations. As Schmid-Petri and colleagues have demonstrated, climate change skepticism is present in both progressive, or liberal, and conservative newspapers. This means, in their view, that those in the climate change denial industry "have found a form of argumentation that advances their goals and is at the same time so subtle that it is covered by all types of media" (Schmid-Petri et al. 2017, p. 508). The effect is to implicitly deny the necessity and legitimacy of fundamental economic and social change to prevent catastrophic human-caused climate change. This, for Schmid-Petri and co-authors, is "a new form of informational bias" (Schmid-Petri et al. 2017, p. 509).

A crucial element of this informational bias is the political normalization of the problem of catastrophic human-caused environment change. By making dispute around climate change a part of normal political debate, in which at least two sides are always available, journalists have made denial possible. Noting what some have called "false balance", Stecula and Merkley argue that journalists "frequently treat topics as debates in which they present 'both sides' in order to adhere to a journalistic norm of objectivity" (2019, p. 6).

Doing so protects journalists from accusations of bias and adds to the drama and conflict that is important to selling the news. But this becomes a problem "when discussing the existence or causes of climate change where the credibility of each side does not have equal weight." By convincing journalists that they have a case, those in the climate denial industry require their views to be covered in 'balanced' reporting. "And, the consequences of this coverage are troubling. Presenting a scientific consensus as a debate confuses the public on the state of the science and, in the case of climate change, possibly reduces support for climate action" (Stecula and Merkley 2019, p. 6).

The usual principles that govern conventional journalism, then, make it more likely to carry the views of those in the climate change denial industry. An imperative to provide 'balance', in particular, leads to the over-representation of the voices of those who are in the climate change denial industry; though the allure of conflict for selling media 'product' ought not to be discounted. But, as Oreskes and Conway pointed out, "the notion of balance … makes sense for political news in a two-party system… But it doesn't reflect the way science works… In an active scientific debate, there can be many sides. But once a scientific issue is closed, there is only one 'side'" (2011, p. 214).

So, even though they speak for a very small number of scientists and do not truly represent another 'side', given the existence of an overwhelming consensus amongst climate science, the views of those in the climate change denial industry continue to be treated with unwarranted respect. Oreskes and Conway noted the findings of Max and Jules Boykoff's study of media stories published between 1988 and 2002. Over half of the stories on climate change published during this period "gave equal time to the majority view among climate scientists as well as to deniers of global warming… Another 35 percent of articles presented the correct majority position among climate scientists, while still giving space to the deniers." Thus "the ideal of balance leads journalists to give minority views more credence than they deserve" (Oreskes and Conway 2011, p. 2150.

Whereas traditional news media organizations may carry the messages of the climate change denial industry as a function of maintaining journalistic standards, online media enable overt forms of climate change denial by bloggers and others who repeat ideas generated within climate change denial think tanks. As Bloomfield and Tillery pointed out, "the Internet and social media serve as platforms for the dissemination of climate change denial and for the construction of virtual denialist identities" (2019, p. 23). Social media, and in particular Facebook, allows "climate change deniers [to] circulate information among themselves and push that information to others." Thus, the Internet offers "supportive environments where skeptical perspectives are communally validated and reinforced." The result is to open "a new rhetorical landscape for the promotion and circulation of (not-so) scientific information, particularly on environmental topics" (Bloomfield and Tillery 2019, p. 23).

These online 'spaces' allow those in the climate change denial industry to bypass the limited constraints imposed by traditional media and to present forms of climate change denial to their audience as though it represented legitimate science. Crucially, this audience includes climate deniers from other forms of media. Thus, in "what Chadwick has termed a 'hybrid media system' that connects online and offline forms of communication in complex ways, … voices… spill over from blogs and Websites to traditional media, above all conservative TV channels like FoxNews" (Schmid-Petri et al. 2017, p. 509). In one image, climate denial "zooms" "through online spaces via the 'climate denial blogosphere,' where it then gets picked up and re-circulated by conservative media." With the goal of those who participate in this blogosphere being "to shift expertise from mainstream science and position themselves as credible scientific resources" (Bloomfield and Tillery 2019, p. 24).

In part, climate change denial reflects the pro-business stance of the personalities, media outlets, parties and other organizations that repeat the messages developed within the climate change denial industry. Addressing climate change is a direct threat to some companies. The most obvious threat being to those involved in fossil-fuel industry, who are the major funders of climate denial think tanks and other organizations engaged in denial. But other industries, such as forestry and agriculture, would be profoundly disrupted by action to address catastrophic human-caused environment change. Such action would also threaten the profitability of many other companies, as it would require investment in new equipment and losses of productivity associated with changing production processes.

But support for climate change denial or a refusal to contradict it and a willingness to offer minor adjustments to address catastrophic human-caused environment change amongst those in conservative parties also reflects their conservatism. The economic and social reforms necessary to address climate change are neither minor nor even modest. So, they do not appeal to those with a conservative disposition, and this is true for both party members and voters. Denying climate change is the easiest way to maintain a conservative approach to policymaking and voting.

Another factor behind members of political parties embracing climate change denial or becoming largely silent on the issue and supporting minor or modest reform is that progressive and radical political parties and movements incorporated addressing catastrophic human-caused environment change in their political agendas. This was and is most obvious in the development of a Green Left, but it can also be found in the eco-anarchist movement. As was discussed in the Introduction to this book, people had already been arguing that fundamental change to society was needed, with many of them arguing for the end of capitalism. Climate change became another reason through which they justified such a change. Denying climate change, then, formed part of a longer-term resistance to calls for radical social and economic change (Stokes 2015). Or, as the then Ronald Reagan Distinguished Visiting Professor at Pepperdine University's Graduate School of Public Policy and former F.K Weyerhaeuser Fellow in Law and Economics at the American Enterprise Institute in Washington DC and senior fellow at the Pacific Research Institute in San Francisco, Steven F. Hayward put it in *Forbes* magazine in 2015:

> Activist liberal elites always need a Grand Cause to satisfy their messianic needs, or for the political equivalent of a dopamine rush. For such people, the only thing worse that catastrophic climate change is the catastrophe of not having a catastrophe to obsess over—and use as an excuse to extend political control over people and resources… (Hayward 2015).

## Part Three: The Climate Change Denial Industry at the Beginning of the Final Hundred Years of Humanity

Even though those in the industry present themselves as a corrective to the Falsehoods perpetrated by almost all climate scientists, this is not the reason that the industry exists. Although the funding for the climate change denial industry is coming increasingly from 'philanthropic' entities such as Donors Trust, its purpose has always been to delay action to address the human causes of catastrophic environment change. The 'science' it offers for those who do not want change is, for the most part, funded by major corporations in the fossil fuel industry.

These corporations have always maintained a (deliberately) complex relationship with those in the climate change denial industry. They are never the sources of climate change denial. They are simply its greatest beneficiaries. Fossil fuel companies have an especially complex relationship with climate science. For in the early years, "some serious science took place in the research facilities of the oil and gas companies, which led them to conclude that climate change was real and dangerous. Nonetheless, they funded... think tanks to spread denial and doubt about the same evidence that their own scientists had produced" (Grasso 2019, p. 109). Thus, at the same time that "ExxonMobil's scientists were contributing to advancing climate science, the company was misleading its stakeholders by spreading doubt about it in its public communications" (Grasso 2019, p. 112).

While the climate change denial industry has been effective in fostering doubt as to the existence and imminent threat of catastrophic human-caused environment change amongst enough people, the spokespeople of major fossil fuel corporations can no longer be seen to publicly dispute mainstream climate science. This required a change in strategy such that these spokespeople must now assure company shareholders and the users of their company's products that they take catastrophic human-caused environment change seriously and are doing everything in their power to reduce and compensate for their contribution to the problem.

Thus, spokespeople for fossil fuel companies are careful to present their companies as taking their 'corporate social responsibility' seriously and being committed to addressing the human causes of catastrophic climate change. Behind the scenes, though, major gas and oil companies continue to support the climate change denial industry, which is where anonymizing donations to the industry is particularly useful. Thus, even though the leaders of these companies may, and do, "openly acknowledge climate science and plan for a less carbon-intensive business, they at the same time 'maintain membership – and in many cases have leadership positions – in trade associations and other industry affiliated groups that spread disinformation about climate science and/or seek to block climate action'" (Grasso 2019, p. 108).

Carroll and colleagues have described this new strategy as "stage 2" denialism. No longer are fossil fuel companies able to contest mainstream climate science, "'stage 1' hard denialism" for Carroll and co-authors. To compensate for this, spokespeople for these corporations have shifted strategy "to propose policies that appear as credible responses to the scientific consensus but do not harm big carbon—the three most typical being greater efficiency in carbon extraction and consumption, new technology, and incremental change inadequate to the scale and urgency of the problem" (Carroll et al. 2018, p. 428).

Thus, for example, the highly influential Canadian Association of Petroleum Producers (CAPP) "proclaims on its website that 'climate change is an important global issue, requiring action across industries and around the globe.' ... Yet CAPP continues to push hard for expanded oil sands production and new pipelines on behalf of its members…" (Klein and Daub 2016). In this later stage of denialism, "the fossil fuel industry and our political leaders assure us that they understand and accept the scientific warnings about climate change — but they are in denial about what this scientific reality means for policy and/or continue to block progress in less visible ways" (Klein and Daub 2016). As Carroll and co-authors put it "stage-2 denial involves talking one way and walking in the opposite direction, by obstructing the kinds of changes that could decarbonize our ecological footprint in a timely manner, but that also threaten corporate profits and control of economic decisions" (2018, p. 446).

For Carroll and co-authors, stage 2 has a specific "architecture" in which "elite cohesion and closure combine with a rich organizational ecology of corporate influence…" (2018, p. 446). In this structure "industry groups, think tanks, advocacy organizations, postsecondary institutions, and research institutes occupy distinct niches in a field of carbon-capital influence that also encompasses aligned sources of finance" (Carroll et al. 2018, p. 446). The benefit of the multiplicity of entities operating in this ecology is that "carbon capital speaks not through a megaphone, but through many voices and from many sites beyond its base in capital accumulation" (Carroll et al. 2018, p. 446).

As the signs of catastrophic human-caused environment change become more evident, the ineffectiveness of the initiatives in which fossil fuel companies have engaged will become increasingly obvious. At this point the deeper constraints on action discussed in Chap. 3 will work to protect fossil fuel industries. Principal amongst them will be the threat to profitability that action to address catastrophic human-caused environment change represents. This goes to the general operation of capitalism and normalcy. And, as Carroll and co-authors pointed out, "corporate influence is, at its core, geared toward protecting investments and profit streams, opening new fields for investment, and minimizing intrusions into profit, such as taxes, regulations, and unions" (Carroll et al. 2018, p. 426). Increasing effort on the parts of those in the climate denial industry, and particularly the corporate spokespeople that channel its ideas, then, will go to "the long game of cultivating a probusiness political and popular culture" (Carroll et al. 2018, p. 426).

Company leaders publicly conceding the existence of catastrophic human-caused environment change and committing their company to addressing it will not damage, but will enhance, their company's reputation. The crucial issues here, of course, relate to processes of accounting, which brings us to the various lies and illusions created by environmental accounting. The System of National Accounts that the UN produced in 1993 triggered attempts, sometimes called Green Accounting, to correct for the usual system of accounting that treats "ecological resources as lying outside economics so that their diminution and degradation are not to be considered economic losses, but merely of 'environmental' consequence" (El Serafy 2015, p.161). Experience with the Triple Bottom Line (sometimes rendered as '3BL') suggests that 'green accounting' will founder on the rock of the requirement for profitability.

For Norman and MacDonald, if "the jargon of 3BL implies that there exists sound methodology for calculating a meaningful and comparable social bottom line, the way there is for the statement of net income, then it is misleading; it is a kind of lie" (2004, p. 256). For them, the 3BL is "good old-fashioned single bottom line plus vague commitments to social and environmental concerns" (Norman and MacDonald 2004, p. 256). Its value is that it allows firms "to make almost no commitment whatsoever" to environmental protection.

3BL allows a firm to "report a number of data points of its own choosing that are potentially relevant to different stakeholder groups – typically in the form of a glossy 3BL report full of platitudinous text and soft-focus photos of happy people and colourful flora" (Norman and MacDonald 2004, p. 256). For without having to calculate a true environmental bottom line "firms do not have to worry about having these 'bottom lines' compared to other firms inside or outside of their sector; nor is there likely to be any great worry about the firm being seen to have declining... environmental 'bottom lines' over the years or under the direction of the current CEO" (Norman and MacDonald 2004, p. 256).

But no matter the means of accounting used to assess the costs of action to address catastrophic human-caused environment change these costs are real. In an article written in 2014, Bullis drew attention to an International Energy Agency report that found that "switching from fossil fuels to low-carbon sources of energy will cost $44 trillion between now and 2050..." and an IPCC finding "that efforts to stabilize levels of greenhouse-gas emissions would require investments of about $13 trillion through 2030" (Bullis 2014).

Both reports were released in 2014 and, as Bullis argued in his article, these estimates "give a sense of what getting away from fossil fuels will cost if we all act now and make smart decisions going forward, and if technologies work out the way we hope they will" (2014). Indeed, the point of Bullis's article was to highlight the fact that delaying action to address the human causes of catastrophic environment change increases the cost of later action. According to Chen and co-authors "if we could have started a global policy in 2010, the cost would be 8% less than if started in 2020 compared to our base policy runs." They concluded that "if we are able to eventually get on an optimal path of emissions abatement toward the 2 °C target, it will likely not be until 2030 or later, further increasing the costs. Our estimate is that delaying the start to 2030 would increase the cost by another 14%" (Chen et al. 2016, p. 16).

And, most importantly, while the economic effects of not addressing the human causes of catastrophic environment change will be devastating in the long run, in the short run the systems of accounting in use will mark addressing those causes as a cost. And even though there are great difficulties associated with the cost of action to address catastrophic human-caused environment change there can be no doubt that it will impose considerable short-term costs. It is at this point that those in the climate denial industry and those who speak for them will rely upon the deeper ideological supports for the preservation of the status quo. For the increasing level of disruption of people's lives required to address the human causes of catastrophic environment change will produce inaction.

## Conclusion

The climate change denial industry is an important part of the politics of the final hundred years of humanity because it represents a means through which specific interests, principally those in fossil fuel industries, can exert influence on popular perception of the existence and level of threat associated with catastrophic human-caused environment change. The most significant feature of this industry is that it has created the impression in enough people's minds that climate scientists are not almost unanimously agree with respect to the reality and catastrophic nature of human-caused environment change.

Either directly, through blogs and other forums, or indirectly, through media personalities and spokespeople who spread their message, the climate change denial industry has been effective in delaying action to address catastrophic human-caused environment change. The result of its efforts is support for parties and politicians who deny the existence or level of threat of catastrophic human-caused environment change. More generally, though, the climate change denial industry has created enough uncertainty to enable the ideological effects discussed in Chap. 3 to produce inaction with respect to address catastrophic human-caused environment change.

The industry has changed its strategies over time and will continue to do so as humanity moves closer to its final hundred years. It began by offering alternative 'science' but has moved to attack climate scientists for ideological bias and self-interest. The delay that the industry has caused and continues to cause increases the costs associated with acting to address the human causes of catastrophic environment change. These costs will continue to grow to the point that they become a principal reason for not acting. This means that the climate change denial industry is moving closer to defending the status quo simply because it is too difficult for society to change.

## References

Båtstrand, S. (2015). More than markets: A comparative study of nine conservative parties on climate change. *Politics & Policy*, *43*(4), 538–561. https://doi.org/10.1111/polp.12122.

Bauer, H. (2017). Science Is Not What You Think. Jefferson (NC): McFarland.

Benestad, R. E., Nuccitelli, D., Lewandowsky, S., Hayhoe, K., Hygen, H. O., Van Dorland, R., & Cook, J. (2016). Learning from mistakes in climate research. *Theoretical and Applied Climatology*, *126*(3–4), 699–703. https://doi.org/10.1007/s00704-015-1597-5.

Bloomfield, E. F., & Tillery, D. (2019). The Circulation of Climate Change Denial Online: Rhetorical and Networking Strategies on Facebook. *Environmental Communication*, *13*(1), 23–34. https://doi.org/10.1080/17524032.2018.1527378.

Brulle, R. J. (2014). Institutionalizing delay: foundation funding and the creation of US climate change counter-movement organizations. *Climatic change*, *122*(4), 681–694. https://doi.org/10.1007/s10584-013-1018-7.

Bullis, K. (2014 May 15). How Much Will It Cost to Solve Climate Change? *MIT Technology Review*, https://www.technologyreview.com/s/527196/how-much-will-it-cost-to-solve-climate-change/. Accessed 15 June 2019.

Carlton, J. S., Perry-Hill, R., Huber, M., & Prokopy, L. S. (2015). The climate change consensus extends beyond climate scientists. Environmental Research Letters, 10(9), 094025. https://doi.org/10.1088/1748-9326/10/9/094025.

Carroll, W., Graham, N., Lang, M. K., Yunker, Z., & McCartney, K. D. (2018). The Corporate Elite and the Architecture of Climate Change Denial: A Network Analysis of Carbon Capital's Reach into Civil Society. *Canadian Review of Sociology/Revue canadienne de sociologie, 55*(3), 425–450. https://doi.org/10.1111/cars.12211.

Chen, Y. H., Babiker, M., Paltsev, S., & Reilly, J. (2016). *Costs of Climate Mitigation Policies*. MIT Joint Program on the Science and Policy of Global Change. http://hdl.handle.net/1721.1/102617. Accessed 22 June 2019.

Cook, J., Nuccitelli, D., Green, S. A., Richardson, M., Winkler, B., Painting, R., … & Skuce, A. (2013). Quantifying the consensus on anthropogenic global warming in the scientific literature. *Environmental research letters, 8*(2), 024024. https://doi.org/10.1088/1748-9326/8/2/024024, 1

Cook, J., Oreskes, N., Doran, P. T., Anderegg, W. R., Verheggen, B., Maibach, E. W., … & Nuccitelli, D. (2016). Consensus on consensus: a synthesis of consensus estimates on human-caused global warming. *Environmental Research Letters, 11*(4), 048002. https://doi.org/10.1088/1748-9326/11/4/048002.

Dunlap, R. E., & Jacques, P. J. (2013). Climate change denial books and conservative think tanks: Exploring the connection. *American Behavioral Scientist, 57*(6), 699–731. https://doi.org/10.1177/0002764213477096.

El Serafy, S. (2015). On the fringe of green accounting—A review article. *Ecological Economics*, (112), 161–163. https://doi.org/10.1016/j.ecolecon.2014.12.005.

Farrell, J. (2016). Corporate funding and ideological polarization about climate change. *Proceedings of the National Academy of Sciences, 113*(1), 92–97. https://doi.org/10.1073/pnas.1509433112.

Farrell, J. (2019). The growth of climate change misinformation in US philanthropy: evidence from natural language processing. *Environmental Research Letters*. https://doi.org/10.1088/1748-9326/aaf939.

Funk, C., & Kennedy, B. (2016). The Politics of Climate. Pew Research Center: Internet. *Science & Tech (October 4)*. Available online at *http://www.pewinternet.org/2016/10/04/the-politics-of-climate*.

Grasso, M. (2019). Oily politics: A critical assessment of the oil and gas industry's contribution to climate change. *Energy Research & Social Science, 50*, 106–115. https://doi.org/10.1016/j.erss.2018.11.017.

Hayward, S.F. (2015 9 June) 12:35pm, Why The Left Needs Climate Change, *Forbes* https://www.forbes.com/sites/stevenhayward/2015/06/09/why-the-left-needs-climate-change/.

Hornsey, M. J., Harris, E. A., & Fielding, K. S. (2018). Relationships among conspiratorial beliefs, conservatism and climate scepticism across nations. *Nature Climate Change, 8*(7), 614. https://doi.org/10.1038/s41558-018-0157-2.

Klein, S., & Daub, S. (2016 September 26). The New Climate Denialism: Time for an Intervention. *Behind the Numbers*. http://behindthenumbers.ca/2016/09/26/new-climate-denialism-time-intervention/. Accessed 22 June 2019.

Leviston, Z., Greenhill, M., & Walker, I. (2015). Australian attitudes to climate change and adaptation: 2010–2014. *Australia: CSIRO*. https://doi.org/10.13140/RG.2.1.2149.0001/1

Lockwood, M. (2018). Right-wing populism and the climate change agenda: exploring the linkages. *Environmental Politics, 27*(4), 712–732. https://doi.org/10.1080/09644016.2018.1458411.

Mann, M. E., & Olson, P. J. (2018 August 20). Keynote Address: The Hockey Stick and the Climate Wars: Dispatches from the Front Lines. *Science Editor*, 41(2), 48–9. https://www.csescienceeditor.org/article/keynote-address-the-hockey-stick-and-the-climate-wars-dispatches-from-the-front-lines/. Accessed 22 June 2019.

Mann, M. E., Bradley, R. S., & Hughes, M. K. (1998). Global-scale temperature patterns and climate forcing over the past six centuries. *Nature*, *392*(6678), 779. https://doi.org/10.1038/33859.

McCright, A. M., Marquart-Pyatt, S. T., Shwom, R. L., Brechin, S. R., & Allen, S. (2016). Ideology, capitalism, and climate: Explaining public views about climate change in the United States. *Energy Research & Social Science*, *21*, 180–189. https://doi.org/10.1016/j.erss.2016.08.003.

Michaels, D. (2019 February 26). The Denial Playbook: How Industries Manipulate Science and Policy from Climate Change to Public Health February 26, 2019. https://naturalresources.house.gov/imo/media/doc/Michaels%20Written%20Testimony%20Denial%20Playbook%20Hearing.pdf.

Norman, W. & MacDonald, C. (2004). Getting to the bottom of 'triple bottom line'. *Business Ethics Quarterly*, 14(2), 243–262. https://doi.org/10.5840/beq200414211.

Oreskes, N., & Conway, E. M. (2011). *Merchants of doubt: How a handful of scientists obscured the truth on issues from tobacco smoke to global warming*. Bloomsbury Publishing USA.

Schaller, S., & Carius, A. (2019). Convenient Truths. https://www.adelphi.de/en/publication/convenient-truths.

Schmid-Petri, H., Adam, S., Schmucki, I., & Häussler, T. (2017). A changing climate of skepticism: The factors shaping climate change coverage in the US press. *Public Understanding of Science*, *26*(4), 498–513. https://doi.org/10.1177/0963662515612276.

Stecula, D. A., & Merkley, E. (2019). Framing Climate Change: Economics, Ideology, and Uncertainty in American News Media Content from 1988 to 2014. *Frontiers in Communication*, *4*, 6. https://doi.org/10.3389/fcomm.2019.00006.

Steentjes, K., Pidgeon, N. F., Poortinga, W., Corner, A. J., Arnold, A., Böhm, G., ... & Sonnberger, M. (2017). European Perceptions of Climate Change (EPCC): Topline findings of a survey conducted in four European countries in 2016. http://orca.cf.ac.uk/id/eprint/98660.

Stokes, B. (2015, November 6). The U.S. isn't the only nation with big partisan divides on climate change. Pew Research Centre. http://pewrsr.ch/1LVyHP4. Accessed 15 June 2019.

The Environics Institute for Survey Research. (2015 October). Focus Canada. Canadian public opinion about climate change, 2015. https://davidsuzuki.org/science-learning-centre-article/focus-canada-2015-canadian-public-opinion-climate-change/. Accessed 22 June 2019.

UT at Austin. (2017). Energy Poll http://www.utenergypoll.com/

Van der Linden, S., Leiserowitz, A., & Maibach, E. W. (2016). Communicating the scientific consensus on human-caused climate change is an effective and depolarizing public engagement strategy: Experimental evidence from a large national replication study. https://doi.org/10.2139/ssrn.2733956.

Wiens, J. A. (2016). *Ecological Challenges and Conservation Conundrums: Essays and Reflections for a Changing World*. John Wiley & Sons.

# Chapter 5
# National Adjustment

This chapter addresses the pressures to which national governments will have to respond due to the effects of catastrophic human-caused environment change on specific communities within their borders. More specifically, the focus is on challenges that arise because of large-scale internal population movements. The first questions that arise here concern the politics associated with the defense of those communities that the members of a government have decided to defend. The following questions relate to the voluntary and forced relocation of those neighborhoods who those in government have decided not to protect from the effects of catastrophic human-caused environment change.

Most of the population movement will result from areas becoming uninhabitable as a result of increased surface temperatures and humidity, desertification and ocean dead zones or from lands no longer being available as a consequence of rising sea levels (accompanied by frequent major storm surges). Increasing toxicity and species depletion will also play a role in rendering areas uninhabitable in the final hundred years of humanity. They will also need to be considered in any examination of the politics of the final hundred years in a specific country or region. Increases in toxicity or species depletion will have profound effects on specific communities in some countries.

The interaction of climate change, species depletion and extinction and rising toxicity will also will be significant in most places. The flooding of the power plant at Fukishima in Japan due to a tsunami is an example of the combined effects of rising sea levels and the release of toxins. That this combination will be repeated in certain countries during the final hundred years of humanity is a real possibility. For many other nuclear power plants are also in vulnerable locations. Nunez estimated that 14 plants in the US are vulnerable to rising sea levels (Nunez 2015). In addition, according to one report, rising sea levels will affect four nuclear power plants and eight sites on which plants have been proposed in the United Kingdom (Edwards 2012). Other sites may be faced with a risk of rising sea levels causing the spread of contamination from nuclear waste dumps (Graham 2017) or other nuclear facilities. Indeed, the highest priority, when it comes to providing for the effects of rising sea



I. Cook, *The Politics of the Final Hundred Years of Humanity (2030-2130)*,
https://doi.org/10.1007/978-981-15-1259-9_5

levels, will be to create sea walls and otherwise protect sites that contain nuclear materials.

Communities will also cease to be viable because their members can no longer support themselves by consuming local plants and animals because of species depletion and extinction and/or changes in migration patterns of species upon which they rely for sustenance. According to Pecl and co-authors, land animals and plants are moving toward the poles at an average of around 17 km per decade and those in oceans are moving at an average of around 72 km per decade (2017, p. 2). This will have profound effects on many communities rendering some unsustainable without considerable support from their government.

The pressure to leave an area may not be simply a result of plant and animal species migrating and becoming extinct, however. This pressure may also come about as a consequence of some species increasing in number. As Pecl and co-authors noted, epidemics that result from the movement of disease carrying insects and animals will threaten communities that had not previously been exposed to these diseases. Populations of the mosquitoes that carry malaria, for example, will spread toward the poles and will affect people living at higher altitudes, for whom it had not previously been a problem (Pecl et al. 2017, p. 3). But the effects of catastrophic human-caused environment change will vary because different species will be affected differently. As Bett and colleagues wrote, "some of the studies… show a positive association between temperature and expansion of the geographical ranges of arthropod vectors (e.g. *Culicoides imicola*, which transmits bluetongue virus) while others are used to illustrate an opposite trend (e.g. tsetse flies that transmit a range of trypanosome parasites in sub-Saharan Africa)" (2017, p. 119).

Another factor that will affect the viability of some communities is that outdoor temperatures and levels of humidity may make it virtually impossible to work in the open air during the day in some places. According to Mora and co-authors, "by 2100, almost three-quarters of the world's human population could be exposed to deadly climatic conditions under high future emissions (RCP 8.5)" (Mora et al. 2017, p. 505). This effect will be profound in tropical areas because they "will be disproportionately exposed to more days with deadly climatic conditions, because these areas have year-round warm temperatures and higher humidity, thus requiring less warming to cross the deadly threshold" (Mora et al. 2017, p. 505). This is likely to be more important at the international level. But those in government in larger countries may find part of their population affected by increased surface temperatures and or humidity.

Thus, even though many of the people in these areas will also suffer from desertification, others may still receive sufficient rain but their locality will simply become too hot and humid for people to work outside. These communities will cease to be viable as their members become less active and productive. If those in government do not come to their aid the members of these communities will be forced to move and they will add to the numbers of internal migrants (or domestic climate change refugees) with which some governments will have to deal. If those in government choose to assist them, then they become yet more claimants for public support in dealing with the effects of catastrophic human-caused environment change during the final hundred years of humanity.

The discussion in this chapter is somewhat general, in that the discussion of national adjustment concerns broader processes that would require tailoring to specific countries or regions for a full account of the politics of national adjustment during the final hundred years of humanity. Different countries will face different mixes of problems. Governments in landlocked countries will not face problems related to the effects of rising sea levels and flooding. So, they will not have to decide whether they will try to protect members of communities who face flooding due to rising sea levels and storm surges or move them to other locations. But those in government in coastal countries will have to address this choice. The greater the numbers of people who live on their coasts the greater the problems some public officials will face (especially that of paying for either assisting the people in these communities to stay where they are or to relocate). Those who govern landlocked countries, however, will have to deal with population effects that result from desertification and increased surface temperatures that make outdoor work virtually impossible.

The people in communities affected by catastrophic human-caused environment change during the final hundred years of humanity will be faced with two choices: to stay where they are and adjust to the effects of catastrophic human-caused environment change or to move to some other part of their country (the politics that results from people seeking to move to other countries will be discussed in the next chapter). Most, if not almost all, of the members of threatened communities will want to stay where they are. They will demand support from others, especially those in government, to do so, as remaining will require investment in alleviating, if not eliminating, the threat or threats they face. Others will accept the impossibility of their situation and accept that moving to another location is necessary. In some cases, the investment necessary to enable people to stay in threatened communities may be too great and they will be required, and in some instances forced, to relocate.

Communities whose members rely on food produced elsewhere may encounter increasing difficulty with respect to a declining capacity to get that food. Global warming threatens the production of wheat, rice, maize, and soybeans, which provide two-thirds of human caloric intake (Zhao et al. 2017, p. 9326). Whereas some areas will become more productive as the global average temperature increases, most will become less productive.

Zhao et al. have calculated the decrease in the productivity of the world's largest producers of wheat, rice, maize, and soybeans for each degree Celsius (1.8 ° Fahrenheit) that the average global temperature increases. In this analysis, maize production declines the most, with the world losing up to 12% of its maize crops with each degree Celsius (1.8° degree F) increase. For wheat the figure was up to 9% of the world's crops being lost for each degree Celsius. Up to 7% of the world's rice crops would be lost for every degree Celsius that the global temperature increased. Soybeans are the least affected, with only 3% of the world's crops being lost when the global temperature increases by one degree (Zhao et al. 2017, p. 9328). Global warming and other factors associated with a decrease in food production will affect every region, so both national and international declines will play roles in some communities becoming decreasingly viable.

Members of coastal communities may be the most threatened during the final hundred years of humanity. As Ramírez and colleagues noted, a combination of climate change and overexploitation as a result of industrial fishing has meant that around 70% of the world's fish stocks have been lost and this has led to profound changes in "fish communities and marine ecosystems since the Second World War" (Ramírez et al. 2017, p. 1).

Muis and co-authors suggested that currently 76 million people are vulnerable to a one-in-a-hundred-year flood due to a sudden surge in sea levels. The countries most vulnerable in their account are China (where 36 million people are vulnerable), The Netherlands (8.4 million – which is half of the population), Vietnam (4.7 million), Egypt (4.3 million), Germany (2.9 million), India (2.5 million), United Kingdom (2.4 million), Japan (2.0 million), Bangladesh (1.4 million) and Belgium (1.1 million) (Muis et al. 2016, p. 7). People in Greenland are also threatened by a sudden rise in the sea level. The number of people threatened does not reach into the millions, as it does with those other countries. But the just-under 20,000 people vulnerable to the effects of catastrophic human-caused environment change on sea levels represent a third of Greenland's population (57,700) (Muis et al. 2016, p. 5).

The viability of coastal communities will also be affected by ocean dead zones, even when they are relatively safe from sea level rise and storm surges. While others will be threatened with all three, the problems of those for whom ocean dead zones are the greater challenge will share something with those who live in areas affected by desertification. Given that coastal areas are generally cooler than inland regions and that other food sources may be available, desertification may well prove a greater challenge for inland communities than those communities that only face ocean dead zones. The point, however, is that communities that are in increasingly unlivable regions will raise important, and difficult, political questions and challenges for those national governments that are still viable as we proceed further into the final hundred years of humanity.

Internal migration and forced removal are not new. Nine suggested that internal displacement affected over 30 million people in 2016, which was twice the number of people who moved to another country. Many displaced people chose to move due to conflicts and disasters that made it dangerous to stay where they were. Government officials decided that remaining was not an option for others, though, and forced them to move to another location (Nine 2016, p. 38). Very different issues arise with respect to voluntary and forced relocation. And these issues do not just affect those who choose or are forced to move. They also affect those who live in regions to which people choose or are forced to relocate.

This chapter is in two parts. The politics of remaining is taken up in the first part. While most, if not almost all, people will prefer to stay where they are, only some will be allowed to stay and have their remaining supported by those in positions of political power. The politics of relocation is the subject of the second part of this chapter. Here the discussion shifts to the rights of people choosing or being forced to move to another location and the responsibilities of those in government to those forced to relocate. This, in turn, requires addressing the rights of those who already inhabit places to which people are forced to remake their homes in another part of the country.

## Part One: The Politics of Remaining

Most, if not all, people will want to stay where they are in the places they know and among those with whom they are familiar. No matter how badly their circumstances are affected by catastrophic human-caused environment change, their first choice will be to stay where they are and maintain their current lifestyles (or as much of them as they can preserve). The uncertainties that arise with respect to relocation will also make remaining their first choice. Most of the people in their country will understand, respect and seek to support that choice. But those who are not forced to move to another part of the country are likely to feel threatened by the possibility that they will have to make room for those forced to move. Helping people to stay where they are will be the first choice, then, for almost everyone in every country. Those who see helping people to remain as an expensive short-term fix that is doomed to fail may well be the only ones who will disagree with treating support for people remaining as the default position.

So, despite the high costs likely to be associated with assisting people to stay where they are this will be the likely first option for everyone. Eventually, though, the costs of supporting people's remaining will be beyond the means of some governments or become too great a burden for those in countries that still have the resources necessary to enable everyone to remain. Helping people to remain in threatened communities will appear to be the correct thing to do until the full costs of doing so becomes clear.

The most practical political question (that is, 'Who pays?') will be an important one. Before that question can arise, however, broader political questions have to be resolved with respect to deciding which communities will be protected from the effects of catastrophic human-caused environment change and which will not. Other broader questions concern the responsibilities that attach to those who are to support those who are to remain where they live during the final hundred years of humanity. In many liberal-democracies, addressing such questions may involve the recognition of a legal right to remain.

In these countries, then, the initial question concerns whether people can assert a right to remain and demand support from their government in exercising that right. People in threatened communities and those who fear that they will have to make room for those in threatened communities will defend the notion of a legal right to remain and support doing what is necessary to enable people to stay where they are. People in both groups will call for recognition of that right by those in governments and in judicial systems. The first set of questions taken up in this chapter concern the basis upon which people in threatened communities could assert a legal right to remain, who might assert it and when they might assert it.

Even if a legal right to remain is not established, though, some people will be provided with the resources to stay where they are. Some communities will be too valuable to be simply abandoned. Many of the world's major cities are situated in coastal areas; as, of course, are all the world's major ports. And the fact that many of the wealthiest and most prestigious suburbs in countries around the world have beach or river frontage cannot be ignored. Some of the most important of the world's mines and gas and oil fields are in areas threatened by desertification and/or extreme temperatures. So it is unlikely that the members of every threatened community will be required to move to another part of the country. This introduces another set of political questions concerning remaining. These include who decides which communities receive support to stay where they are and how that decision is to be made. This is a second set of questions that arise with respect to the politics of remaining.

Supporting a community's remaining, either because of a general right to remain or the need to protect communities that are considered too valuable to be abandoned, imposes considerable costs that leads to the most basic political question: "Who pays?" While relocation is expensive, doing what needs to be done to allow people to stay where they are may prove even more expensive. Calculating how much would have to be invested to enable the members of a community to stay where they are is difficult. Rights may be an important starting point for a discussion of the politics of remaining in many countries. But cost is going to be another factor that will be considered. Over time, it is likely to determine which communities will have their localities protected from all the effects of catastrophic human-caused environment change and which will not.

A legal right to remain will be particularly important for communities whose members maintain traditional lifestyles, as they are likely to be most affected by a combination of increased surface temperatures, ocean dead zones and species depletion (which includes changes in migration patterns that mean that certain species are no longer available for hunting or fishing). While the loss of their traditional food sources is significant, the disruption of their spiritual practices may be even more deeply disturbing for those who have maintained traditional lifestyles. Finland's Skolt Sámi, for example, are struggling to help the Atlantic salmon to survive and to maintain their spiritual connection to the species. While in Eastern Siberia the Chukchi are facing the end of their traditional nomadic reindeer-herding practices (Pecl et al. 2017, p. 3). Helping these communities to stay where they are may represent a moral imperative. To lose a source of food is one thing. To lose a spiritual anchor is another. For the over-riding principle is that being displaced from their traditional lands would devastate those seeking to live a traditional lifestyle.

Another factor that must be acknowledged in this context is that the experience with displacement has not been positive. As a 1998 report from the UN Commission on Human Rights suggested, relocating people "breaks up families, cuts social and cultural ties, terminates dependable employment relationships, disrupts educational opportunities, denies access to such vital necessities as food, shelter and medicine, and exposes innocent persons to such acts of violence as attacks on camps, disappearances and rape" (Deng 1998). Relocation is particularly damaging for poor people because they can no longer rely on each other for support, as they had been able to do in their original community's location. The core principle in this context is that people will take their social problems with them. Relocation is also particularly hard for people in rural communities, as their skills are often better fitted to where they were than to where they are moved (Nine 2016, pp. 38–39).

A desire to stay where one is may be so strong that some people will go to any length to do so. The government of the Maldives, for example, constructed an artificial island as part of its strategy to allow its people to remain, though this raises a variety of issues with respect to international law (Gagain 2012). Recent studies show that even the members of communities badly affected by catastrophic human-caused environment change want to stay where they are. In 2017, for example, members of small communities in the Philippines refused to relocate despite the fact that they were exposed to threats of flooding as a result of land subsidence. They preferred to seek to reduce the effects of flooding, while maintaining the option of a temporary move to the mainland during extreme flooding. Limited resources meant that they had not been able to introduce the best flood reduction strategies available. But they showed a dogged determination to remain where they were (Jamero et al. 2017, p. 585).

## *The Politics of Remaining I: Rights of Remaining?*

In many countries, the first questions that must be addressed with respect to communities remaining where they are despite the effects of catastrophic human-caused environment change concern whether the people in these communities can assert a legal right to remain. Their right to stay where they are would seem obvious. But forced evacuations in the face of disaster do happen and this might not be so simple if children's lives are threatened by a decision to stay where they are. Further, if there *is* a legal right to remain, then it would probably be an individual right. But a community could become unviable if enough individuals choose to move to another part of the country. So, the right to remain might be better understood as a communal right. This could mean that enough members of a community would have to commit to remaining for the members of that community to assert a collective legal right to remain. This leads to the question of the level of support a community whose members wanted to stay where they are could legitimately expect from their government.

In liberal-democracies at least, the starting-point for a discussion of the legal rights that operate when the effects of catastrophic human-caused change threaten the viability of communities is to assume that people have a right to remain because the members of those communities want to stay where they are. First and foremost, because a legal right to freedom of movement, which includes the freedom not to move, is generally considered a basic human right in liberal-democracies. But, as a second source, so are property rights, and this can be understood as a right to remain on ones property (McDermott and Gibbons 2017, pp. 591–2). McDermott and Gibbons, though, are convinced that a right to remain can be derived from international human rights conventions and court decisions (McDermott and Gibbons 2017, pp. 590–1).

Nine has argued that the existence of a legal right to remain relates to people's right to stay in their home. She disagreed with those, such as Moore (2015) and Stitz (2013), whose arguments implied that the right to remain was about a legal right to stay in a specific region. Instead she argued that people's right to remain was a right to remain in their homes. Our connection to our homes is, for Nine, even more important that our connection to the region where we live. Our homes, in this account, are so important to who we are as individuals and what we are trying to do with our lives that a legal right to remain is an individual right to remain in our homes (Nine 2016, pp. 50–1).

In the final hundred years of humanity many, perhaps most, people whose communities are threatened by the effects of catastrophic human-caused environment change will assert a right to remain. The way that they establish this right will vary. The people with the means to do so will demand, either individually or in class actions, that the courts recognize this right. Some will fail at the outset, as the existence of a right to remain will have to be asserted through a local legal system and only some of them may allow such a right. Others will lack the financial resources to use the legal system to establish a right to remain.

Even when citizens establish such a right, however, international law recognizes that government officials can decide to move people against their wishes in some circumstances. The questions that arise in this context concern how those in government could establish a power to deny, or override, a legal right to remain on the parts of those whose communities are threatened by the effects of catastrophic human-caused environment change. The principles that apply when those in government seek to deny or override a person or community's right to remain will have to be established during the final hundred years of humanity.

This leads to a further reason for believing that the starting-point for approaching national adjustment should be with a right to remain. Not starting with a right to remain could mean that those in government would not fully explore and properly assess the steps that would have to be taken to allow people to stay where they are (Ferris 2014, p. 6). This point must be related to the likelihood that those in governments tend to pursue options that retain their political support. This means that they will be tempted to choose to defend a right to remain on the parts of their supporters and to move communities threatened by the effects of catastrophic human-caused environment change if they do not contain a significant number of their supporters.

For various reasons, then, people will seek to establish a right to remain. They will point to their connection to their homes or their regions and their concerns with respect to the places to which they are to be moved. Their next move will be to call upon government officials to help them to stay. Recognition of a right to remain by those in government will not be enough. The final phase is to force those in government to provide support necessary to enable their right to remain.

Government officials will resist the recognition of this extension or application of the right to remain on two grounds. The first is that they determine that the costs outweigh the benefits associated with defending some of the regions that have become uninhabitable due to catastrophic human-caused environment change. The other reason not to recognize a right to remain is that to do so might create an ongoing obligation to defend the members of a community from the continued, and increasing, effects of catastrophic human-caused environment change.

Predicting the costs associated with continuing to defend communities from the ongoing effects of catastrophic human-caused environment change is likely to prove difficult given the increasing levels of environmental degradation. Once protected areas are created, however, people will invest resources in those areas. This will increase pressure on those in government to provide further protection and make it harder for those in government to relocate people when the time comes (Hino et al. 2017, p. 364). Governments in countries in which people enjoy extensive civil rights and a legal system through which to assert rights to protection are likely to be subjected to court challenges in which residents with the resources to do so seek to establish a right to protection. And if such a right is recognized then ongoing responsibilities with respect to protection will also have to be satisfied. Even when a right to remain has not been established each time those in governments determine that a community is too valuable to abandon and requires protection it may be difficult for them to stop providing protection even as costs mount (Hino et al. 2017, p. 365).

The question of how decisions to defend some places might be made will be addressed in the following section. Here the questions are more general and concern whether a legal right to remain applies in all circumstances and at any cost. And the answer to this question comes down to whether the legal right to remain is taken to be a limited or an unlimited right. This a particularly important question here because the costs associated with compensating for the effects of catastrophic human-caused environment change will increase as conditions worsen during the final hundred years of humanity.

The issue of whether a right is a limited or an unlimited right has arisen in a variety of contexts. The closest analogy here may be the right to healthcare for the elderly. The aging of the population in most countries has combined with improving medical technology to place financial demands on governments that most are struggling to meet. Few Treasury officials would be as direct as the then Japanese Finance Minister, Taro Aso, was when he told old people that they should hurry up and die (Nelson 2013). But, even though meeting the growing costs associated with providing healthcare for an aging population are not as challenging as they are for the Japanese government, those in government in most countries would understand Aso's response to the budgetary pressure that he faced (Nelson 2013).

The right to health is not, of course, age dependent and The Office of the United Nations High Commissioner for Human Rights has stated that "a country's difficult financial situation does NOT absolve it from having to take action to realize the right to health" (OUNHCHR 2008). Indeed, governments are supposed to devote a maximum amount of the resources available to guaranteeing the right to health irrespective of the age of the person to whom those resources are directed. Yet those in the OUNHCHR acknowledge that the level of support that those in government have an obligation to provide should reflect the resources that are available to them. This suggests that the right to health is unlimited in theory but is limited in practice. The OUNHCHR simply sidesteps the question of whether investing in promoting the health of some people is appropriate.

Daniel Callahan is a somewhat lonely, and often reviled, figure in continuing to support the rationing of health care for the elderly (an argument he has made for over 30 years) (Callahan 1987). He argues for the introduction of careful procedures for making decisions on when an elderly patient should not be given medical treatment. Many doctors are making the decision for their governments by simply withdrawing or reducing the support they provide, and governments are supporting them, albeit tacitly, in doing so. But no formal process exists to stop doctors from continuing to provide support if their patients demand it. The legal right to medical treatment is officially an unlimited right, but sometimes unofficially a limited right.

Those who claim a legal right to remain will present it as an unlimited right. Those in government will claim that it is a limited right. Their principal argument for limiting this right will be with respect to the cost of guaranteeing an ongoing right to remain for the members of all communities. It cannot be left to local representatives to make unofficial decisions to limit this right, in the way that the decision to withhold treatment for elderly patients is left to doctors. In addition to the inequity and uncertainty this creates, limiting the right to remain is likely to be a vote loser. Until, that is, voters who do not reside in areas made inhospitable to human occupation due to catastrophic human-caused environment change begin to resist the substantial tax increases necessary to fund infrastructure projects from which they will get no direct benefit. But this raises the more practical issues dealt with in the next section.

The question of whether the right to remain is a limited or an unlimited right is related to another complicating factor in this context: the possibility of "managed retreat" for the members of those communities threatened by rising sea levels (Hino et al. 2017, pp. 364). Managed retreat raises a variety of legal, logistical and, ultimately, political questions that add complexity to an already difficult terrain. The first question that arises in the context, though, concerns the use of the term 'managed retreat'. Members of "some communities prefer … to avoid the negative connotations of 'retreat'" and prefer "planned relocation" or "managed realignment" (Dannenberg et al. 2019, p. 2). Whichever terms is preferred, operationalizing 'managed retreat' in the context of catastrophic human-caused environment change will present a variety of problems with respect to principle and practice.

Those who do not live in liberal democracies will have to negotiate questions associated with the protection of some communities and their members staying where they are, when others will be forced, by nature or government, to move. Much depends on the judicial systems in liberal-democracies, though those in legislatures (parliaments or congresses) in these countries may seek to extinguish any legal right to remain. One of the most important of the questions that arise, at least in liberal-democracies, is whether, if a legal right to remain is recognized, it is a limited or unlimited right.

## *The Politics of Remaining II: The Practicalities of Remaining?*

Remaining comes at a cost; but so does relocation. Before the discussion moves to address questions that relate to the cost of enabling people to stay where they are, then, it is worth noting Hino's calculation (in 2017 US$) that the upfront costs of relocation are between 10,000 and 100,000 dollars per person (Hino et al. 2017, p. 368). The decision is not, then, between a government supporting cost-free relocation and refusing costly remaining. Further, some communities will be protected even if the cost of staying far exceeds the cost of relocation. For, as has been noted, some communities are too important economically or symbolically for them to be abandoned due to the effects of catastrophic human-caused environment change. While the politics of remaining raises questions of political principle, it is also a pragmatic politics of cost-benefit analyses (which take political costs, and not only financial or economic costs, into account).

If courts do not force or if they cannot continue to compel those in government to guarantee a right to remain for all people, then decisions will have to be made about which communities will be protected. Economic and political factors will drive these decisions, though political factors will be stronger than economic ones. Political factors will give larger communities an advantage (and not just in democracies).

Cities, then, will be more likely to be protected from the consequences of catastrophic human-caused environment change than towns or villages, with major cities having an added advantage in any competition over who will be protected from effects that would otherwise force them from their homes and communities. More voters live in major cities and many major cities are national or state/provincial capitals. Major cities are also important economically, as they are usually industrial and financial centers. Thus, as Dannenberg and co-authors suggested, "large cities are more likely to have the political influence needed to secure national funding for disaster response" (Dannenberg et al. 2019, p. 10).

Major coastal cities will have the greatest advantage in the competition over which communities will get to stay where they are despite the effects of catastrophic human-caused environment change. They have large populations, are likely to be administrative centers for governments and businesses and usually have significant port facilities. Many also have symbolic value for the people of the nation. Most of the world's major cities are near the coast, so the protection they will require will be against rising sea levels, storm surges and extreme rainfall events. But all port cities will have an advantage when it comes to gaining protection from the effects of catastrophic human-caused environment change. Ports that mainly transport fossil fuels, however, will be among those vulnerable to losing support for their protection during the final hundred years of humanity.

The picture is somewhat different for communities outside of major cities. Their economic importance, though, will result in the members of some rural and regional communities being supported in staying where they are despite the effects of catastrophic human-caused environment change. Agricultural regions are often economically important and the protection of the communities in those regions from desertification is likely to be considered a national priority. The economic value of mining and other communities engaged in resource extraction industries will also mean than many will be supported in remaining.

Their lower population density, however, makes communities in regional and rural areas vulnerable to losing in a competition over which communities will be protected from the effects of catastrophic human-caused environment change to political factors. The symbolic value of some of these communities may be great, however, and the question of whether a community will be supported will not be a simple one. Electoral systems may prove important in this context, as some can amplify the 'voice' of those living outside cities. The chances or staying where they live for those who live in communities in rural and regional areas will be greatly enhanced if they benefit from a voting system that gives their votes greater weight. The problem that people in these communities will face is that desertification will reduce their productivity and their economic importance will decline.

If they cannot use the court system to force government officials to provide what they need to remain the members of inland communities will generally find it harder to stay where they are. Unless they benefit from changed wind patterns, temperatures will increase, making it harder for people to work outdoors during daylight hours. Prolonged droughts will combine with extreme storms that result in greater water runoff to create problems with irrigation. Communities in most inland regions will struggle to resist the effects of desertification unless they receive heavy investment in water storage facilities. But they will not require protection from rising sea levels and extreme storm surges. So, protecting them may prove less expensive, and therefore more attractive, to those in government.

Competition will intensity between communities seeking government protection from the effects of catastrophic human-caused environment change. The general principle, of course, is that, unless measures are taken to curb their power, communities whose members wield significant financial or political resources will have a greater chance of being able to obtain government support for staying where they are.

But this competition may also affect the party systems in many countries. As the problems that the members of specific communities face grow in significance the pressures associated with representing regional interests will affect major political parties in many countries. The cohesion of such parties will come under increasing strain as the leaders of these parties seek to develop national policies with respect to protecting a variety of threatened communities. Pressures leading to division and eventual fragmentation of major parties are likely to increase. Smaller parties that represent specific regions are likely.

It is possible that the various parties that form from a fragmented major party might focus on the representation of the interests of specific types of threatened community. If coastal zones contain around half of the world's population, as Yan and co-authors suggested (2016, p. 480), then policies directed to addressing the common interests of people in coastal areas might provide a party with an electoral advantage. Parties representing regional agricultural interests are present in many countries and could adapt to the new political environment. But they too may well come under pressure to divide as the needs of those in agricultural communities diverge as they deal with different changes to the environment. If this is the case, forming or maintaining a party that represents those in all regional agricultural interests may be difficult.

Religious factors may decide which communities will receive protection from the effects of catastrophic human-caused environment change. Religious discrimination is a fact of life in most countries (Finke et al. 2017). When the members of a dominated religious group, usually a minority religion, are concentrated in a specific area, their government may be less responsive to their needs. The reverse also applies, with governments being more responsive to the needs of people who live in areas occupied by those of the dominant faith.

The politics of the final hundred years of humanity will require the development of methods for determining which communities will be protected from the effects of catastrophic human-caused environment change and enabled to stay where they are. Communities will deploy whatever resources they have available to them, including economic, political and symbolic resources, in an increasingly desperate competition for government support. Party systems will be disrupted in most countries and different forms of regional politics and even micro-regional political parties may result from the competition amongst communities over protection from the effects of catastrophic human-caused environment change.

## *The Politics of Paying for Remaining*

It is impossible to generalize about the economic costs associated with protecting specific communities from the effects of catastrophic human-caused environment change. Most of the research into costs has been done with respect to sea walls. Thus, there are few guides to specifying the costs of adaptation strategies with respect to desertification and life-threatening surface temperatures and humidity (Semenov and Shewry 2011). The principal difficulty in this context is that a variety of measures may have to be introduced to deal with problems associated with desertification and life-threatening surface temperatures and humidity.

To support communities whose members live in areas in which it is impossible to work outdoors during sunlight hours will require major infrastructure investments. While maintaining productivity in areas affected by desertification requires, amongst other things: diversifying and changing the crop species being cultivated or livestock being raised (Nakashima and Suenaga 2017), introducing water-saving irrigation technologies (Alva et al. 2017), creating more effective systems for capturing water (Kim et al. 2017) and increased cultivation in greenhouses (Van Cotthem 2015).

Very few communities will have the resources to provide for their own protection from the consequences of catastrophic human-caused environment change. Most will only be able to stay where they live if they receive support from their government. People who live on coasts in well-populated areas are often wealthier and it is possible that, in some cases, they could pool their resources to provide for their own protection. There may also be some mountain communities whose members might be able to fund their remaining (but these are forested areas and their vulnerability to fire might make their situation harder to defend than those who live on the coasts).

Of course, those with economic resources are generally able to extract resources from government officials. So, people in wealthier communities will usually receive government assistance in providing for their protection from the effects of catastrophic human-caused environment change. The greater likelihood, then, is that communities with the resources to do so will invest in making the protection of their region the highest political priority. The result will be to justify government investment in ensuring that they can stay where they are.

The subtitle of, then well-known American political scientist, Harold Lasswell's 1936 work sums it up: *Politics: Who Gets What, When, How*. In this work, Laswell adopted the science of politics approach, discussed in the Chap. 2, in which those who studied politics merely described how political order was created through and in the face of competition for resources. The politics of the final hundred years of humanity will simply provide another site for "the study of influence and the influential" and "who get what there is to get." The difference with respect to this period is that most attention will go to the third of the three things that Laswell argued people are trying to get: "deference, income, safety" (with the elite getting the greater share than the mass because others defer to them) (Lasswell 1958, p. 295).

One aspect of the politics of the final hundred years of humanity that will continue from the preceding period, then, is that the interests of 'the influential' will be best served. Economic importance, as was pointed out in the preceding discussion, will provide the most persuasive grounds for government investment. But members of the elite will also have the resources to use the media and the court system to compel those in government to contribute to the protection of the places in which they want to stay where they are.

Those in governments will be likely to seek to force people in wealthier communities to contribute to their remaining, however, particularly as the costs associated with protecting communities increase. Governments in some countries simply will not be able to afford the resources to support remaining for many, if any, communities. But even those in government in richer nations have both finite resources and a need to avoid being seen to favor one community over another. So, other than in exceptional circumstances, such as protecting capital cities and communities that are economically significant, government officials will seek to force much of the burden onto those who can pay for their protection from the full effects of catastrophic human-caused environment change. This will prove an important aspect of the politics of remaining during the last hundred years of humanity.

Those who have been refused government support to help them to stay where they live despite the effects of catastrophic human-caused environment change will contest their government spending money to enable the members of other communities to stay where they are. They will use whatever political and economic resources they have to challenge policies that mean that some people get to remain stay where they are when they cannot. They will use principles, such as 'user pays' and 'small government' policies, to challenge the legitimacy of government action to enable others to remain. In addition, or if this fails, they will take those in government to court to prevent them from spending public resources to enable what those who mount such a challenge will claim to be private gains.

In the process of protecting industry in some places, and to appease aggrieved members in unprotected communities, those in government may also seek to ensure that greater amounts of the money generated in protected communities will be recouped by government agencies. The possibility of nationalizing industries in protected areas will have to be addressed. In part, because this will ensure that more money is available to invest in protection. In part, because it will reassure members of unprotected communities, who will be forced to leave their homes, that a collective benefit results from protecting those in other communities and allowing them to keep their homes. The current ideological hegemony of neoliberalism will have to be fully displaced, however, before 'nationalization' policies can be implemented.

In many countries, the first political questions to be addressed with respect to national adjustment will concern which communities will receive public support in enabling their members to stay where they are by protecting them from the effects of catastrophic human-caused environment change. In liberal-democracies, this will initially be approached in terms of whether a legal right to remain can be established and the obligations of those in government when it comes to people seeking to enforce that right. If they can, then the next question concerns whether this is a limited or an unlimited right.

But important political questions will arise even in those countries in which individual rights are not recognized or a right to remain is not recognized. There is little likelihood that governments in many countries can protect all communities from the effects of catastrophic human-caused environment change. But some communities will be protected because of their economic importance or their members have significant political resources. Even when people can claim a right to remain they will face resistance from those in government with respect to public funds being used to pay the full costs for protecting communities and this will become a limited right. Extended political and/or legal battles will characterize the final hundred years of humanity in most countries. Even the most oppressive of governments in countries that have no cultural tradition of acknowledging human rights will not escape people's struggles to stay where they live during the last hundred years of humanity.

Most governments, though, have not developed effective and comprehensive principles for deciding whether to protect or to abandon a community. Governments in some countries, though, are starting to address the problem. A few have begun to develop processes for deciding which communities to protect and which to sacrifice to the effects of catastrophic human-caused environment change. The government of the Federated States of Micronesia, for example, "has proactively adopted a managed-retreat strategy to redirect development inland." While the UK's Shoreline Management Plans "have identified locations where the current 'Hold the Line' strategy will no longer be the preferred strategy in the future; retreat – either managed by a government agency or not – is likely to take place" (Hino et al. 2017, p. 368).

None of the decisions as to which communities are to be protected and which will be exposed to the full effects of catastrophic human-caused environment change will be easy. The stakes involved and the various political resources mobilized during the making of these decisions will ensure that they are fraught with difficulty.

The possibility that those in government might lose support from members of communities whose towns and cities are not protected from the effects of catastrophic human-caused environment change will also mean an ever-present threat of fragmentation and a possible decent into intra-national conflict in some countries.

## Part Two: The Politics of Relocation

While increases in sea level will not be continuous, and will occur in rapid bursts, they will not be immediately life threatening. Nor will desertification and ocean dead zones directly threaten people's lives. Droughts are another matter, though, while flooding because of storm surges and extreme rainfall events may pose immediate risks. But these threats will not be permanent. Over time, however, recognition of the inevitability of relocation may cause members of a community to decide that they want to move and, in some cases, many of them will seek government support to do so. One reason that they will need government support is that, in many cases, they will not be able to sell their fixed assets to settle elsewhere. When they can sell them the value of many, if not all, of their fixed assets (their homes, investment properties and businesses) will be greatly reduced by the time they accept that they must move to another part of the country.

Even when they choose to move to another part of the country, the people who relocate due to effects of catastrophic human-caused environment change will be in a similar position to people being forced to move to another part of the country. As has been noted, staying where they are will be the preferred option for most people in threatened communities and for most of those in government. But long-term projections could lead public officials to decide to move the members of a community before they face a direct threat to their lives, so forced relocation may still happen. The principle that public officials have a duty to protect people's lives that overrides people's right to remain is recognized in international human rights law (McDermott and Gibbons 2017, 590). The United Nations Covenant on Economic Social and Cultural Rights, for example, allows forced evictions in exceptional circumstances (CESCR 1997).

People in most liberal-democracies have rights when it comes to forced relocation that have been established by courts or through legislation and their public officials also have legally defined responsibilities in such circumstances. It has been clearly established, however, that those in government can ignore or over-ride human rights when it comes to forcing people to move to another part of the country due to natural disasters ((McDermott and Gibbons 2017, p. 588). They may have to be ignored or over-ridden when it comes to the forced relocation of threatened communities during the early period of the final hundred years of humanity. And such rights may no longer exist in later stages and forced relocation will be seen as a luxury available in less desperate times. Once again, in most countries increased attention will have to be paid to human rights during the final hundred years of humanity. This is particularly important, as some of these rights will not be fully respected in particular cases.

One question that arises in this context concerns the rights of those who are force to move to another part of the country and, in particular, what relocated people ought to expect that they will be provided with in the places to which they have been moved. Governments may seek to encourage voluntary relocation and to do so will have to make relocation attractive by providing high levels of support for those who move to another part of the country. But resources will become increasingly limited during the final one hundred years of humanity and those in government may want to control the expectations of those who choose or are forced to relocate.

Other questions concern whether conditions ought to be the same for those who have decided to move and for those who have been forced to move. Members of a community will probably find it easier to have their rights respected and to claim higher levels of government support when they have been forced to move. But this does not require that their rights should be different when it comes to the place to which they have been moved. One reason that their rights and the support they can claim ought to be the same is that the members of threatened communities will never choose to move to another part of the country if they believe that they will be penalized for choosing to relocate rather than waiting to be forced to move.

It may be strategic, as has been noted, for public officials to provide better conditions for members of threatened communities early in the last hundred years of humanity in order to induce people to choose to move to another part of the country earlier rather than later. So, a deliberate policy of discriminating in favor of those willing to relocate early by providing them with better conditions than those who choose to move late may be sound policy. For people to wait until the very last moment to move might mean that too many people in too many communities suddenly require support to move to another part of the country.

This is likely to be the situation in the middle period of the final hundred years of humanity. By then it will be too late to worry about responsibilities to communities threatened by the effects of catastrophic human-caused environment change. Those in governments will be focused on the much bigger problems of continuing to be viable and ensuring the survival of as many of their people as possible. They will also be working with more limited resources, given that they will already have been investing in protecting some communities and moving others. So, the question of whether those who move to another part of the country early ought to be treated in the same manner as those who do so late is one for the early years of the final one hundred years of humanity.

## *Establishing Rights in the Context of Forced Relocation Due to a Political Decision Not to Protect a Community*

In theory at least, if they abide by the law in doing so, members of most nations are free to move to other parts of the territory over which their government has legitimate authority. Being forced to move to another part of the country due to a political decision not to protect a community and choosing to relocate because of the effects of catastrophic human-caused environment change are not the same, though. If there are any principles related to a general right to move to another part of the country, they do not help in addressing questions concerning the rights of the people forced to move to another part of the country because public officials have decided not to protect their community.

Questions concerning the rights of those forced to relocate are only some of those that must be faced during the final hundred years of humanity. Additional questions concern the rights of those who already live where the members of a threatened community are moved. One of these questions relates to whether their rights are the same when the people who move to another part of the country have chosen to move and when those who are relocated have been forced to do so. Taking in those who have chosen to move to another part of the country seems a different thing from taking in those who have been forced to relocate. This may make attempts to constitute the category of domestic 'climate change refugee' as important as attempts to constitute a category of international 'climate change refugees.' The former may meet less resistance than the resistance that the latter has met (which will be discussed in the next chapter).

The main problem faced by people forced to move to another part of the country during the final hundred years of humanity is that the value of any assets that are in the place that they are leaving will fall significantly, in most cases to zero or close to zero. Getting out earlier rather than later, certainly before 2030, may well be the best strategy. But, as was discussed in the first part of this chapter, the connection people have to their homes and neighborhoods will make it hard for them to make an early choice to relocate. And the time at which it is possible to sell up and move may well have passed once the effects of catastrophic human-caused environment change become more evident.

Some people will have assets that are not located in an area to which they choose or are forced to move to another part of the country. Most will not. Many of the principles taken up in the following discussion relate to people whose assets are solely or principally located in the area that they are leaving. But one question that arises for people who possess assets that will not lose all of their value as a result of catastrophic human-caused environment change relates to whether a means test will apply with respect to the support with which they will be provided (and whether such a test applies at all or in the same way to those who have chosen to relocate as to those being forced to do so). A means test makes some sense in this context, especially when people have chosen to move to another part of the country. But means tests have their own challenges (See Gustman et al. (2017) and Samwick (2018)).

This assumes, of course, that governments have the capacity to provide support for the members of community who are forced to move to another part of the country because of a political decision not to protect them from the effects of catastrophic human-caused environment change. The resources will simply not be available in many countries. But it is also likely that some communities will not be given support even when they are forced to move. In many cases, those in government will not be able to provide support because they have exhausted their resources through protecting those communities thought too valuable to relocate and through other expenditure required to address the effects of catastrophic human-caused environment change.

Significant social tensions will result from relocations forced by political decisions not to protect the members of some communities from the effects of catastrophic human-caused environment change. They will be intensified in poorer countries, as people are forced to move to another part of the country without government support and people in the places to which they relocate are forced to deal with an increased population without government support. This is a recipe for conflict and creates the possibility of forms of violent confrontation and humanitarian crisis.

If the necessary resources are available, politicians and parties will find it difficult to retain support from the people if they do not commit to providing support for the members of communities forced to relocate due to a political decision not to protect them from the effects of catastrophic human-caused environment change. The only condition under which this might arise is if the people in a threatened community belong to an ethnic or religious minority who can be neglected without significant political cost.

Two sets of issues, which give rise to questions of legal rights, apply to being forcibly relocated. The first set concern the places to which people are relocated. The second relate to the conditions people can expect in the places to which they are moved. When addressing these issues, it is important to bear in mind both that the relocations result from a political decision not to protect a community and that most people will have to rely on government support for their relocation.

A preliminary, and more general, question concerns how much control governments will exert over relocation. One possibility is that those in governments will provide a fixed sum of money to everyone forced to move to another part of the country to allow them to choose the place to which they will move. This may well be supplemented by payments to the local authorities that will be required to provide infrastructure for the additional community members for whom they will have to provide. In federations, payments may also be made to regional or state governments. This will tend to produce relocation to places with greater employment prospects for those who rely on working for their income. It will also produce relocation to places where it is cheaper to live for those who do not rely on working for a living. The problem that arises if this approach is taken is that too many people may choose to move to the same area. But the approach will have some political benefits in liberal-democracies, in which an individualist political culture prevails.

Another possibility is that those in government will provide inducements to move to specific regions by providing greater sums of money for relocation to places to which they want people to move to another part of the country. Increasing the population of places that require protection from the effects of catastrophic human-caused environment change seems an unlikely strategy (unless this proved politically popular, cheaper or more efficient). Attracting people to places that do not require protection, in the early years of the final hundred years of humanity at least, is likely to be the more attractive option for public officials.

A third option is for those in government to create accommodation and to compel people to move to that accommodation as one part of a political decision not to protect their community from the effects of catastrophic human-caused environment change. One of the questions that will arise if public officials take up this option concerns whether all the members of a community are to be relocated to the same area or whether the decision as to the area to which they move will fall to family units or individuals.

Those in governments will generally find it easier to deal with large groups when it comes to relocation. So, they will prefer to work with communities, rather than cater to the desires of families or individuals. Even providing several options with respect to where communities might relocate and allowing their members to choose could easily lead to over-subscription with respect to some places and under-subscription for others. Avoiding this outcome will require detailed prior planning and careful work with the members of all communities forced to move to another part of the country because of a political decision not to protect them from the effects of catastrophic human-caused environment change and compensate for the effects they will have in the communities to which they will be relocated. It is likely, though, that forced relocation will occur in waves and that, no matter how much planning is done, those in the earlier waves will have more desirable places to relocate available to them. This will leave those in later waves to face having to move to less desirable areas. This will give rise to resentment amongst those in later waves of relocation and this sense of grievance will make it harder for them to adjust to their new circumstances.

There is a danger that a romantic view of communities will be adopted and that communities will be treated as if each member is in harmony with every other member of the community. Government planners, as noted above, will find working to relocate communities easier than working with families or individuals. But the rise of neoliberal ideology means that those in government in some countries will be predisposed to approach relocation due to catastrophic environment change as a problem of addressing the interests of isolated individuals with no ties to other members of the community.

As Taylor Aitken and coauthors suggested, "Neoliberal societies seem to prize the experience of the individual above the collective. Community activism and belonging can therefore be countercultural, as are sustainable development aspirations to collective action" (Taylor Aiken et al. 2017, p. 3). Those in government in countries with a more individualistic culture will find it harder to get acceptance for a community-based approach from people who see themselves, first and foremost, as individuals. On the other hand, those in government in countries whose culture emphasizes the important of attachment to community for identity will be liable to adopt a community-based approach to forced relocation. But, in these cases, relocation will often have profound costs due to the connection between community and place.

Most of the principles that have been developed with respect to forced migration concern eviction or displacement during a natural disaster. While there are differences, a political decision not to support a community made unviable by the effects of human-caused catastrophic environment change is related to politically forced eviction and the displacement of a community as a result of a natural disaster. So, the principles that have been developed with respect to both eviction and displacement during a natural disaster provide some guidance with respect to the political processes necessary to reduce the number and intensity of conflicts that will result from forced relocation.

Paragraphs 16 and 17 of the UN Covenant on Economic Social and Cultural Rights, which concern eviction, make it clear that informing the people and allowing them to engage in discussions concerning their forced relocation will help to legitimize forced relocation (UN). This is reflected in Principle 7(d) of the United Nations Office for the Coordination Of Humanitarian Affairs' 'Guiding Principles on Internal Displacement': that people being forced to move to another part of the country should be provided with full information about their relocation and the compensation they will receive. Principle 7(f) directs attention to the need to involve the people being moved, especially women, in the planning and management of their relocation (UNOCHA 1998).

Ensuring an acceptable standard of living for those forced to move to another part of the country will be vital to ensuring that forced relocation will not produce unmanageable social conflict. Each country and culture will have a different conception of adequate housing and the infrastructure and services that are necessary for an acceptable standard of living. But the basic requirements include "accessibility, … cultural adequacy, habitability, legal security of tenure, suitability of location, and access to essential services such as education and healthcare" (McDermott and Gibbons 2017, p. 592). Access to employment is yet another important factor in whether proper provision has been made for people forced to relocate due to the effects of catastrophic human-caused environment change.

## *Compensating Those Who Already Occupy the Place to Which Others Are Moved*

But attention cannot be directed solely to those forced to move to another part of the country due to the effects of catastrophic human-caused environment change. Not everyone who lives in the place to which others are forced to move will welcome them. People forced to relocate are likely to face significant discrimination from the people who already live in the places to which they have been relocated. Laws that prohibit some, generally more overt, forms of discrimination may be in place. But these laws may not be respected, and they do address covert forms of discrimination. Indeed, people forced to move to another part of the country may encounter significant levels of discrimination from local officials who lack the resources to cater to their needs or simply do not want to do so.

The effect of forced relocation on a local economy will vary and may well affect some locals more than it does others. But all will be affected. A study of the effects of internal migrants in Indonesia found that each 1% increase in the number of immigrants resulted in close to a 1% decrease in the average income of people living in the place to which the immigrants moved. Locals who competed with migrants for jobs that required little education and skills were the hardest hit in terms of both wages and rates of employment (Kleeman and Magruder 2017).

Supporting those in government in local and state or regional authorities to induce them to accept the people being relocated to the areas they control will be important for preventing discrimination and conflict. People forced to move to another part of the country are much more likely to experience resentment and discrimination from people who already live in the places to which they move if resources are not provided to facilitate their acceptance. Ethnic or religious discrimination may mean that those forced to move to another part of the country will encounter discrimination no matter what support is offered to people who live in the places to which they move to another part of the country. This will be more extreme, though, if people who are forced to relocate are perceived to be an economic burden by those who live in the places to which they relocate.

The struggle of internal migrants in Mumbai and Kolkata serve as examples of the sorts of problems faced by those forced to move to another part of the country as a result of a political decision not to support a community made unviable as a result of the effects of catastrophic human-caused environment change. The first form of discrimination they face is that internal migrants to Mumbai and Kolkata have to establish their status as citizens by providing legal documentation. Without this documentation they cannot claim their rights to housing and food rations. The discrimination that internal migrants face in Mumbai is a result of ethnic discrimination. In Kolkata it is a result of discrimination against the poor. According to Abbas, discrimination "shapes citizenship outcomes, not administrative capacity issues that are tied to underdevelopment." Further, Abbas also noted that, while the Indian Constitution provided citizens with rights, "national policies are not sufficient for guaranteeing minority rights" (Abbas 2016, p. 164).

The problems faced by internal migrants in India illustrate the problems that will be faced by people forced to move to another part of the country as a result of a political decision not to protect their community from the effects of catastrophic human-caused environment change. These problems may be reduced because those forced to relocate might expect some sympathy from those forced to accept them into their towns or cities. But this sympathy will disappear quickly if the people affected by others relocating to their areas are not involved in decisions concerning this relocation and bear economic costs for which they are not compensated.

Just as those forced to move to another part of the country have to be informed about and included in the decision-making processes associated with their relocation, so too must the people who live in the places to which people will move be informed and involved. Information and inclusion will not resolve all of the tensions that will have to be managed during the final hundred years of humanity. Those forced to relocate by government officials who have decided not to protect them from the effects of catastrophic human-caused environment change already have grounds for grievance. Just as those forced to accept them have reasons to resent both those outsiders and the government officials who have forced them to accept outsiders compelled to abandon their homes and neighborhoods. There is much political work to be done in this context.

Relocation forced by a political decision not to protect communities from the effects of catastrophic human-caused environment change creates different types of political question. One type relates to the rights of those forced to move to another part of the country. Another set relates to the places to which they will relocate and the conditions that they will encounter in those places. The final type of questions concern recognition of the rights of and support for the people who already live in the places to which others will be forced to move.

How these questions will be answered will depend on the political culture in which those in government operate and through which they will approach the process of relocation. Providing a lump sum payment to individuals, or families, forced to move to another part of the country and asking them to decide to where they will move, for example, is an option that is more likely in liberal-democracies where individual freedom is paramount. But it is likely to result in difficulties in administration and a tendency for significant population increase in more desirable areas. Those in authoritarian governments, on the other hand, are less likely to be constrained by liberal-individualism and able to direct people to the areas to which they want them to go.

An important difference emerging from this example is with respect to the effect on policy of the way that those in government think of their people (and the people think of themselves). 'Community' is a crucial social unit in some cultures and for them approaching relocation as the relocation of communities will make considerable sense. Working with larger aggregates is also attractive from the perspective of policy makers and those who implement policy. Such approaches represent what Arnall, Hilson and McKinnon characterized as a 'top-down' approach to "the governance of climate displacement and resettlement [that] is often technical, managerial and depoliticized." Top-down approaches are often "implemented in a manner that is 'thin' and 'tokenistic'. Formal participation schemes are always initiated from 'outside' and the intervening agencies ultimately retain control over the moments and the manner of participation" (Arnall et al. 2019, p. 668).

Arnall, Hilson and McKinnon wrote of a variety of dangers associated with "treating communities as homogenous entities whose claims always genuinely represent the interests of community members" (2019, p. 669). Creating places and forcing people or communities to relocate to those places is also the best approach from the perspective of those who must plan for relocation to places already set up for it. Treating a community "as a relatively discreet spatial unit, with a more or less homogenous social structure and a set of shared norms or understandings… considerably simplifies the resettlement process, as communities are viewed as bounded by a single administrational rule and unproblematically represented by local leaders or elites" (Arnall et al. 2019, p. 669). And at least some aspects of this aspect of the politics of the final hundred years of humanity may well be determined by which approach is easiest to design and implement and not by the rights, needs or interests of those affected by relocation.

## Conclusion

For the individuals affected, determining the best strategy to adopt for people living in threatened communities during the final hundred years of humanity depends on a variety of factors. How long they can be protected may be the most important. But how important their location and community are to their nation's economy will be a very important factor. How much it costs to protect them will also be significant. While most people in most communities will want to stay where they are, they would be unwise if they did not consider leaving early in the final hundred years of humanity. To wait too long will leave them vulnerable to having to move to another part of the country without the support of and assistance from their government.

While the conditions of people forced to relocate within national borders will vary from country to country, similar problems and questions arise. The first concern the basic question of whether people have legal rights to remain and what rights can be claimed by those forced to relocate. Whether people affected by catastrophic human-caused environment change can demand support from their government for staying where they live will be important, then. As will the level of support they will receive to help them to do so.

While there may not be a right to be relocated, other questions concerning rights arise for those affected by forced relocation. Then a myriad of practical questions arises concerning how remaining and relocation will be approached and decisions made. The goal of this chapter was not to resolve or provide guidelines for resolving the difficult questions related to national adjustment to the effects of catastrophic human-caused environment change. It was simply to outline some of the questions of principle and practice that arise in this context.

## References

Abbas, R. (2016). Internal migration and citizenship in India. *Journal of Ethnic and Migration Studies*, *42*(1), 150–168. https://doi.org/10.1080/1369183X.2015.1100067.

Alva, A., Albaho, M., Kumar, G., Annabi, M., Stevens, W., & Dodge, K. (2017). Agricultural Water Use and Technologies for Adaptations to Water Deficits. In *Water Resources in Arid Areas: The Way Forward* (pp. 501-521). Springer, Cham. https://doi.org/10.1007/978-3-319-51856-5_29

Arnall, A., Hilson, C., & McKinnon, C. (2019). Climate displacement and resettlement: the importance of claims-making 'from below'. *Climate Policy*, 1–7. https://doi.org/10.1080/14693062.2019.1570065

Bett, B., Kiunga, P., Gachohi, J., Sindato, C., Mbotha, D., Robinson, T., … & Grace, D. (2017). Effects of climate change on the occurrence and distribution of livestock diseases. *Preventive veterinary medicine*, *137*, 119–129.https://doi.org/10.1016/j.prevetmed.2016.11.019.

Callahan, D. (1987). *Setting Limits: Medical Goals in an Aging Society*. New York: Simon & Schuster.

CESCR (Committee on Economic, Social and Cultural Rights). (1997). Sixteenth session – General comment No. 7: The right to adequate housing: Forced evictions, Comment 2. https://www.un.org/ruleoflaw/files/Housing%20and%20Property%20Restitution.pdf. Accessed 22 June 2019.

Dannenberg, A. L., Frumkin, H., Hess, J. J., & Ebi, K. L. (2019). Managed retreat as a strategy for climate change adaptation in small communities: public health implications. *Climatic Change*, 1–14. https://doi.org/10.1007/s10584-019-02382-0.

Deng, F. M., (1998, February 11). Guiding Principles on Internal Displacement, United Nations E/CN.4/1998/53/Add, *Commission on Human Rights*, http://www.un-documents.net/gpid.htm. Accessed 25 June 2019.

Edwards, R. (2012, Mar 7). UK nuclear sites at risk of flooding, report shows, *The Guardian*. https://www.theguardian.com/environment/2012/mar/07/uk-nuclear-risk-flooding. Accessed 25 June 2019.

Ferris, E. (2014, March 14). Planned Relocations, Disasters And Climate Change: Consolidating Good Practices And Preparing For The Future (Background Document Sanremo Consultation, 12–14 March) UNHCR. http://www.unhcr.org/53c4d6f99.pdf. Accessed 25 June 2019.

Finke, R., Martin, R. R., & Fox, J. (2017). Explaining discrimination against religious minorities. *Politics and Religion*, *10*(2), 389–416. https://doi.org/10.1017/S1755048317000037.

Gagain, M. (2012). Climate change, sea level rise, and artificial islands: Saving the Maldives' statehood and maritime claims through the constitution of the oceans. *Colo. J. Int'l Envtl. L. & Pol'y*, *23*, 77–120. https://heinonline.org/hol-cgi-bin/get_pdf.cgi?handle=hein.journals/colenvlp23&section=6.

Graham, K. (2017, Nov 30). Rising sea levels threatens U.S. nuclear waste dump in Pacific, *Digital Journal*. http://www.digitaljournal.com/news/environment/usa-nuclear-waste-dump-threatens-to-contaminate-pacific-ocean/article/508854. Accessed 25 June 2019.

Gustman, A. L., Steinmeier, T. L., & Tabatabai, N. (2017). Means testing social security: Income versus wealth. *National Tax Journal*, *70*(1), 111–132. https://doi.org/10.17310/ntj.2017.1.04.

Hino, M., Field, C. B., & Mach, K. J. (2017). Managed retreat as a response to natural hazard risk. *Nature Climate Change*, *7*(5), 364. https://doi.org/10.1038/NCLIMATE3252.

Jamero, M. L., Onuki, M., Esteban, M., Billones-Sensano, X. K., Tan, N., Nellas, A., … & Valenzuela, V. P. (2017). Small-island communities in the Philippines prefer local measures to relocation in response to sea-level rise. *Nature Climate Change*, *7*(8), 581–6. https://doi.org/10.1038/NCLIMATE3344.

Kim, H., Yang, S., Rao, S. R., Narayanan, S., Kapustin, E. A., Furukawa, H., … & Wang, E. N. (2017). Water harvesting from air with metal-organic frameworks powered by natural sunlight. *Science*, *356*(6336), 430–434. https://doi.org/10.1126/science.aam8743.

Kleemans, M., & Magruder, J. (2017). Labour market responses to immigration: Evidence from internal migration driven by weather shocks. *The Economic Journal*, *128*(613), 2032–2065. https://doi.org/10.1111/ecoj.12510.

Lasswell, H. (1958). *Politics: who gets what, when, how*. Cleveland, Ohio: World Pub. Co.

McDermott, R., & Gibbons, P. (2017). Human rights and proactive displacement: determining the appropriate balance between the duty to protect and the right to remain. *Disasters*, *41*(3), 587–605. https://doi.org/10.1111/disa.12217.

Moore, M. (2015). *A political theory of territory*. Oxford University Press.

Mora, C., Dousset, B., Caldwell, I. R., Powell, F. E., Geronimo, R. C., Bielecki, C. R., … & Lucas, M. P. (2017). Global risk of deadly heat. *Nature Climate Change*, *7*(7), 501. https://doi.org/10.1038/NCLIMATE3322.

Muis, S., Verlaan, M., Winsemius, H. C., Aerts, J. C., & Ward, P. J. (2016). A global reanalysis of storm surges and extreme sea levels. *Nature communications*, *7*, 11969. https://doi.org/10.1038/ncomms11969.

Nakashima, K., & Suenaga, K. (2017). Toward the genetic improvement of drought tolerance in crops. *Japan Agricultural Research Quarterly: JARQ*, *51*(1), 1–10. https://doi.org/10.6090/jarq.51.1.

Nelson, R. (2013, Jan. 23). Japanese minister wants old people to "hurry up and die" *Time*, http://newsfeed.time.com/2013/01/23/japanese-minister-wants-old-people-to-hurry-up-and-die/. Accessed 25 June 2019.

Nine, C. (2016). Water crisis adaptation: Defending a strong right against displacement from the home. *Res Publica*, *22*(1), 37–52. https://doi.org/10.1007/s11158-015-9310-1.

Nunez, C. (2015, December 16). As Sea Levels Rise, Are Coastal Nuclear Plants Ready? *National Geographic*. https://news.nationalgeographic.com/energy/2015/12/151215-as-sea-levels-rise-are-coastal-nuclear-plants-ready/. Accessed 25 June 2019.

OUNHCHR (Office of the United Nations High Commissioner for Human Rights) (2008). The right to health - Fact Sheet No. 31, http://www.ohchr.org/Documents/Publications/Factsheet31.pdf

Pecl, G. T., Araújo, M. B., Bell, J. D., Blanchard, J., Bonebrake, T. C., Chen, I. C., … & Falconi, L. (2017). Biodiversity redistribution under climate change: Impacts on ecosystems and human well-being. *Science*, *355*(6332), eaai9214. https://doi.org/10.1126/science.aai9214.

Ramírez, F., Afán, I., Davis, L. S., & Chiaradia, A. (2017). Climate impacts on global hot spots of marine biodiversity. *Science Advances*, *3*(2), e1601198. https://doi.org/10.1126/sciadv.1601198.

Samwick, A. A. (2018). Means Testing Federal Health Entitlement Benefits. *Tax Policy and the Economy*, *32*(1), 173–210. https://doi.org/10.1086/697141.

Semenov, M. A., & Shewry, P. R. (2011). Modelling predicts that heat stress, not drought, will increase vulnerability of wheat in Europe. *Scientific reports*, *1*, 66. https://doi.org/10.1038/srep00066.

Stilz, A. (2013). Occupancy rights and the wrong of removal. *Philosophy & Public Affairs*, *41*(4), 324–356. https://doi.org/10.1111/papa.12018.

Taylor Aiken, G., Middlemiss, L., Sallu, S., & Hauxwell-Baldwin, R. (2017). Researching climate change and community in neoliberal contexts: an emerging critical approach. *Wiley Interdisciplinary Reviews: Climate Change*, *8*(4), e463. https://doi.org/10.1002/wcc.463, 3.

UN, International Covenant on Economic, Social and Cultural Rights http://www.ohchr.org/EN/ProfessionalInterest/Pages/CESCR.aspx. Accessed 25 June 2019.

UNOCHA (United Nations Office For The Coordination Of Humanitarian Affairs). (1998, August). *OCHA Guiding Principles on Internal Displacement* https://docs.unocha.org/sites/dms/Documents/GuidingPrinciplesDispl.pdf. Accessed 25 June 2019.

Van Cotthem, W. (2015, December 3). Building living tunnel greenhouses for rural people in the drylands. *Desertification* https://desertification.wordpress.com/2015/01/18/living-greenhouses-for-the-drylands/

Yan, X. H., Boyer, T., Trenberth, K., Karl, T. R., Xie, S. P., Nieves, V., … & Roemmich, D. (2016). The global warming hiatus: Slowdown or redistribution?. *Earth's Future*, *4*(11), 472–482.

Zhao, C., Liu, B., Piao, S., Wang, X., Lobell, D. B., Huang, Y., … & Durand, J. L. (2017). Temperature increase reduces global yields of major crops in four independent estimates. *Proceedings of the National Academy of Sciences*, *114*(35), 9326–9331. https://doi.org/10.1073/pnas.1701762114.

# Chapter 6
# Civil Wars and International Conflicts

While the problem of catastrophic human-caused environment change is an international problem, no international agency exists to address it and its consequences. Nor, as the failure of the various climate agreements or accords indicates, does a will exist that might result in meaningful international collective action to maintain global political and social order in the face of catastrophic human-caused environment change. The problem, though, is not simply that no agency exists to impose changes that would prevent catastrophic human-caused environment change and divert from a path leading to the final hundred years of humanity. The problem is also that no international agency exists that might prevent the different forms of conflict (civil war, resource conflicts and border conflicts) that will emerge and/or become more intense during the final hundred years of humanity. Despite the good intentions of many of those involved in the United Nations, the organization remains a vehicle for the expression of national interest.

Conca responded to moves on the parts of officials from a number of governments to construe catastrophic human-caused environment change as a security issue in 2017 and 2018 by asking whether there was "a climate role for the most powerful and contentious of all UN organs, the Security Council" (Conca 2019, p. 6). This is a controversial move and several concerns have been raised with respect to any role for the UN Security council in addressing catastrophic human-caused environment change and preventing the onset of the final hundred years of humanity. For some leaders the problem is that involving the Security Council would disrupt the work being done by those who participated in the UN Framework Convention on Climate Change. "Others reject the 'securitization' of what they see as primarily an issue of international law and sustainable development. Still others, wary of the Council's concentrated powers and unaccountability to the full membership, oppose stretching its mandate beyond traditional matters of war and peace" (Conca 2019, p. 6).

The main problem that all UN agencies face, though, is that people and their governments will become more inward looking as they seek to protect themselves from the effects of catastrophic human-caused environment change. This means that



I. Cook, *The Politics of the Final Hundred Years of Humanity (2030-2130)*,
https://doi.org/10.1007/978-981-15-1259-9_6

they will be decreasingly likely to commit resources and troops to supporting any attempts by UN agencies to address the human causes of catastrophic environment change, to broker peace or to otherwise prevent the conflict to which catastrophic human-caused environment change will contribute.

While some reject the notion of 'climate change refugees,' as will be discussed below, increasing numbers of people will seek refuge in other countries as the effects of catastrophic human-caused environment change intensify. The influence that the UN and other agencies have exerted with respect to the acceptance of other types of refugees has relied thus far on moral pressure and shaming, as there is no power to enforce Conventions associated with the acceptance of those seeking refuge. Wodak (2019) might be right to suggest that a post-shame era now enables resistance to claims by refuges. As the effects of catastrophic human-caused environment change intensify, however, whatever moral pressure and shaming once functioned to support claims by refugees will become less effective. This is because the effects of catastrophic human-caused environment change on the people of their countries will force national leaders to exclude others to provide for their own people.

The previous chapter was an account of the adjustments that those in position of political power will be forced to make due to the pressures created by catastrophic human-caused environment change. Few governments will have the capacity to make those adjustments for all their people. Others will be able to support some of their people but will not be able to extend their support across all their territory. Many, in the end most, national governments will collapse due to catastrophic human-caused environment change. This will result in millions of people attempting to migrate to countries with stable governments and sufficient resources to cope with an influx of immigrants. According to Heslin and colleagues, "predictions of future movements of people from climate change and environmental causes vary from 25 million to 1 billion in 2050" (Heslin et al. 2019, p. 242). This will be the principal source of conflict during the final hundred years of humanity. But catastrophic human-caused environment change will also result in resource conflicts, as those in national governments work to secure water, food and livable land for their citizens at the expense of people of other nations.

Civil wars and international conflicts will also mark the final hundred years of humanity. In many cases, this will be the result of conflict over diminishing resources. Civil wars will break out in those nations in which members of different ethnic, religious or cultural groups within the same nation compete for increasingly scarce goods. Most of the international resource conflicts during the final hundred years of humanity will concern access to water. But conflict over other resources, such as over fishing grounds, will also occur. The other type of international conflicts will occur at the borders of nations whose governments and people have the means to adjust to the effects of catastrophic human-caused environment change.

These border conflict will not be between nations as much as they will be between those in viable nations and those who want to migrate to those nations because their governments have not survived the effects of catastrophic human-caused environment change. Some small island nations, for example, will be so badly affected by rising sea levels and storm surges that they may have to be abandoned. The 'borders' referred to in this discussion are not necessarily the borders that nation-states currently occupy. Those in government in countries with extensive land masses may, and probably will, decide to abandon specific regions and will, effectively, change their borders.

While many resist media representations of "'sinking' or 'titanic' places that helpless islanders need to evacuate as 'climate refugees,'" as Klöck and Nunn suggested, the image remains potent. Small islands are "popularly characterized as 'canaries in the coalmine' or 'barometers of climate change,' being among the first and worst affected landmasses" (2019, p. 197). Other nations will not be viable because they are too badly affected by desertification, temperature increase that result in unlivable conditions and the spread of disease.

These countries will be vulnerable to civil war in the first instance, with migration leading to border conflicts being an effect of civil war (though rising sea levels that flood island nations will be another cause of the migration that will lead to border conflicts). Civil war is likely to be the main form of conflict in the early decades of the final hundred years of humanity. Care is required in this context, though, because both internal divisions and resources are necessary for civil wars to break out and be sustained. Catastrophic human-caused environment change will not cause war directly, however. It will exacerbate existing social tensions such that they become violent conflicts that tear nations apart.

Eventually, then, even those in viable nations will find themselves in conflict over resources that have become scarce because of catastrophic human-caused environment change. The first of these conflicts will concern access to water. But conflicts will also result when fishing industries are affected by ocean acidification, species depletion and changes in the migration patterns of fish stocks due to changing sea temperatures. International conflict over resources may be limited in the early period of the final hundred years of humanity. But access to resources will become an increasing source of conflict between nations later in the final hundred years of humanity.

Livable lands will be the main resource over which conflicts will occur in the later years of the final hundred years of humanity. This will be a continuation of border conflict in the northern hemisphere, as only a limited part of the Arctic region has not been claimed and the governments of the countries that border the area (the US, Russia, Canada, Denmark and Norway) will be quick to claim it. The Antarctic, however, represents a much larger area of unclaimed territory and presents a more complicated picture when it comes to the forms of inter-governmental conflict that will occur during the final hundred years of humanity.

Of the three forms of international conflict discussed here, resource conflict will be closest to conventional warfare. Battles over the control of water or other resources will pit military against military. Border conflicts will be more severe and brutal than those previously experienced. While we are familiar with the conflicts that result when people seeking to migrate are detained at national borders, many more people will be forced to migrate due to of the effects of catastrophic human-caused environment change. And those in governments struggling to survive the effects of catastrophic human-caused environment change will be less constrained by international protocols and are likely to treat those on their borders much more harshly.

Civil war will be principal source of border conflict and is the first form of conflict dealt with in this chapter. It then moves to consider civil wars, resource conflicts and border conflicts promoted and intensified by catastrophic human-caused environment change. However, preliminary discussion is required with respect to the idea of 'climate wars.' While others have rejected the idea for very good reasons, as will be explained below, the global situation may become so difficult during the final hundred years of humanity that it may make sense to talk of 'climate wars' (both with respect to civil wars and international conflicts). At this stage, however, referring to climate change triggered conflicts makes more sense.

## Climate Wars

The early-2000s marked the beginning of a series of reports, books, journal articles, blogs, documentaries and news stories about 'climate wars'. In many cases the wars being written about were civil wars. In other cases, though, the 'climate wars' being referred to were between nations. In many of these accounts, droughts, floods, desertification, rising sea levels, the spread of disease and the other effects of catastrophic human-caused environment change lead to migration and international tensions that result in civil wars and international conflicts.

The attraction of the phrase 'climate wars' is obvious. Alternatives, such as 'climate change triggered conflicts' or 'climate change heightened conflicts' lack the simplicity and drama of 'climate wars'. Using these more accurate alternatives does not have the same effect on people as using the phrase 'climate wars.' And many of those who used the phrase 'climate wars' were seeking to highlight the urgency of action to address the human-causes of catastrophic environment change.

As early as 2003, the authors of a report commissioned by the US Department of Defence, warned of "skirmishes, battles, and even war due to resource constraints" caused by climate change (Schwartz and Randall 2003, p. 2). Think tanks and consultants were quick to take up the idea of 'climate wars.' Nicholas Stern, who wrote *The Review on the Economics of Climate Change* for the UK Government in 2006, spoke of "an extended world war" (Tertrais 2011). Journal articles and books about 'climate wars' started appearing later in the 2000s. Dyer's *Climate Wars: The Fight for Survival as the World Overheats* came out in 2008 (Dyer 2008), as did Welzer's *Climate Wars: What People Will Be Killed for in the 21st Century* (which was first published in German in 2008 and in English in 2012 (Welzer 2012).

Warnings of impending 'climate wars' have continued. A 2016 publication in *Scientific American* suggested "the U.S. military is focusing on 2 hotspots where climate change could lead to new conflicts – sub-Saharan Africa and the Asia-Pacific region. It is carefully watching a 3rd, the Arctic. A 4th theater, the Middle East, could also be on the list; but the U.S. Central Command is currently preoccupied with ongoing conflicts there in Syria, Iraq, Yemen and Afghanistan" (Holland 2016, p. 62).

The problem with the use of the phrase 'climate wars' is that it projects a simple causal relationship between climate change and war. Buhaug summarized the problem with such a characterization admirably when he pointed out that this implies that political and military organizations will "resort to force in an organized and coordinated fashion only because temperature heats up or rainfall comes in unexpected ways." (2016, p. 333). Buhaug was not alone in challenging the idea of 'climate wars'. Froese and Schilling noted that "neo-Malthusian-simplified links between violent conflict and environmental deprivation… have been widely criticized…". More importantly, they added that "the community of researchers is divided on whether links between climate and conflict might be drawn or rejected" (Froese and Schilling 2019, p. 27).

Conca also observed that many scholars have tried "to capture the effects of climatic conditions or changes on violent conflict…" but that the "findings to date have produced as much contention as consensus among scholars. Profound disagreements endure over the relationship between, say, drought and violence, or whether there is evidence for a climate signature in the Syrian civil war" (Conca 2019, p. 6).

Buhaug probably goes too far in suggesting that the term 'climate wars' "should be avoided at all costs" (2016, p. 335). But the position that climate change simply and automatically produces civil war or international conflict is not valid. If nothing else, this leaves out important political factors, such as governmental capacity and the level of division in a country, which play crucial roles in whether difficult situations descend into violence. Thus, as Froese and Schilling suggested, "climate change and resource scarcity are [better]… perceived as multipliers of risks and threats that exacerbate existing societal problems such as poverty, injustice, social insecurity, violence, terrorism, or civil war" (Froese and Schilling 2019, p. 24).

Perhaps the only theory of a direct causal relationship between catastrophic human-caused environment change and conflict is that of heat stress causing violent conflict. Miles-Novelo and Anderson noted research showing "that uncomfortably warm temperatures increase aggressive thoughts and feelings, or what we call irritability" (2019, pp. 2–3). They argued for the presence of "a slower indirect effect that takes many years to develop" in which "the likelihood of a child developing into a violence-prone adult will be increased by rapid climate change" (Miles-Novelo and Anderson 2019, p. 5). They concluded "that rapid climate change has already played an essential role in violent behavior at the individual and intergroup levels" and that it "will play an increasingly significant role in the near future as climate change effects become more severe…" (Miles-Novelo and Anderson 2019, p. 7).

Even in this context, however, the relationship between climate change and violence is not that of simple, direct causation. For the most part, violence will increase as "known risk factors for a developing fetus or child eventually becoming a violence-prone adult will become more prevalent because of climate change–induced ecological disasters" (Miles-Novelo and Anderson 2019, p. 5). The violence that occurs is because "climate change effects exacerbate other pre-existing political/social/economic risk factors that are known to have large effects" (Miles-Novelo and Anderson 2019, p. 7).

## Civil Wars

The first civil war widely claimed to have been caused by catastrophic human-caused environment change began in the Darfur Region of Sudan in 2003. As Selby and Hoffmann wrote, "Sudan generally and Darfur specifically are often held up as providing paradigm-defining evidence of our looming future of climate change-induced conflicts" (Selby and Hoffmann 2014, p. 360). Once again, positing a direct causal link between climate change and civil war is overly simplistic and misses important factors that function as crucial links between them.

As Olsson pointed out with respect to the Darfur conflict, "three decades of drought in the area can at least partially explain the observed disintegration of markets and the subsequent rise of social tensions" (Olsson 2016, p. 1). This indicates the form of the causal chain that will be crucial in determining whether and where civil wars break out during the final hundred years of humanity: prolonged drought, which can be linked to climate change, led to the collapse of markets. This exacerbated existing social tensions, some of which may be related to catastrophic human-caused environment change, and produced violence that escalated to civil war (Raleigh et al. 2015, p. 196). In like manner, Raleigh, Choi and Kniveton examined various conflicts in Africa and argued that increases in food prices, which was linked to decreased rainfall and droughts, was a significant cause of the violence that occurred. (2015, p. 196).

The collapse of a market does not simply result from economic factors, though. As Selby and Hoffmann argued, "political economic dynamics, rather than changes in resource availability, have been the main proximate determinants of environment-related conflict in the Sudans" (Selby and Hoffmann 2014, p. 367). This is also why Raleigh, Choi and Kniveton recommended political interventions to manage prices to sustain local markets as the best way to prevent outbreaks of violence. They also noted that the effect of food price and climate on "political violence is largely determined by the goals and coordination abilities of affected groups, who incorporate the experiences of marginalization and hardship into their respective conflict strategies." (Raleigh et al. 2015, p. 187).

The civil war in Syria, which began in 2011 and is all but over in 2019, has also been taken to have resulted from climate change. Former US Vice President Al Gore's second climate change documentary *An Inconvenient Sequel: Truth to Power* goes further than this to suggest a link between climate change and terrorism: an extraordinary drought affected food production, which "led to increased conflict, a civil war, mass exodus, global conflict over immigration and, as a consequence, the emergence of Islamist terrorist group ISIS." This according to Mann, expressed

Gore's "genius for joining the dots in the global mapping of climate impacts" (2017, p. 400).

But Gore's, Mann's and other such accounts of the Syrian war are, according to Selby and co-authors, part of "a long, sad history of people making overblown claims about climate change and conflict" (2017, p. 244). In the case of climate change and the war in Syria, they added, the problem of these simplistic accounts "has been accentuated by the readiness of certain natural scientists… to feed this… policy and public discourse while using evidence casually and failing to engage with relevant social scientific research" (Selby et al. 2017, p. 255).

Selby goes even further in a later article to argue that the drought in northeast Syria in the mid-2000s was a result of politics and not climate change. He rejected "the thesis that a climate change-induced drought in northeast Syria sparked local livelihood insecurity and local out-migration, in turn contributing to civil war onset…". Against "environmental security perspectives [that] imagine a causal sequence which runs essentially from environmental changes to social and then political effects," he wrote, "it is shown here that even in this exemplary case the causality is much closer to the reverse: a case of politically-induced environmental scarcity and vulnerability" (Selby 2018, p. 2).

In Syria the trajectory toward civil war began, according to Able and co-authors, with "growing water scarcity and frequent droughts, coupled with poor water management, [which] led to multiyear crop failures, economic deterioration and consequently mass migration of rural families to urban areas" (Abel et al. 2019, p. 239). The rapidly growing urban population resulted in "overcrowding, unemployment and increased inequality [that] put pressure on urban centers and finally contributed to the breakout of political unrest". Thus, catastrophic human-caused environment change's effect is to aggravate "such drivers of conflicts as poverty, food insecurity and inequalities" (Abel et al. 2019, p. 239).

Selby's criticisms reveal two problems with simplistic causal scenarios in which political effects follow environmental effects and never precede them. First, these scenarios underplay the important contribution that politics makes to the outbreak of war. And, second, the lack of care that is manifested in these scenarios creates the "risk of climate conflict talk fuelling climate scepticism" (Abel et al. 2019, p. 239).

The latter may well reflect the fact that the intention behind the work of some of those who reject a link between catastrophic human-caused environment change and war is to promote skepticism with respect to climate change. Gleick argued that a problem with Selby et al.'s 2017 article was that it failed "to consistently distinguish between 'causality' and 'influence' related to the role of climate change in the severe drought in the region and the subsequent social and political unrest." Further, according to Gleick, "while the authors note in a few places that the research studies they critique do not typically claim that climate change 'caused' the Syrian unrest, they themselves regularly repeat that very argument as a strawman that they then try to debunk" (Gleick 2017, p. 249).

While both problems are important, the need to recognize the importance of political processes in whether climate stress leads to war is more important. Climate change creates stress that means that problems like poverty, social tension and corrupt or ineffective government may result in violence and in some cases war. But, as Shaffer has pointed out, it is possible for people "to cooperate in response to climate stress without resorting to violence" (Shaffer 2017, p. 224). Cooperation will become increasingly hard to achieve as the pressures caused by catastrophic human-caused environment change increase, however.

But, while cooperation remains possible, we can assume that climate change leads, albeit not directly, to conflict and war. The path to the final hundred years of humanity is replete with poor, but understandable, decisions. And people and their leaders are more likely to make poor decisions when the effects of catastrophic human-caused climate change increase social tensions and result, through a complex causal chain, in the demise of already weak governments.

Yet another problem with the assumption that catastrophic human-caused environment change simply causes civil war, as Salehyan and Hendrix noted, is that violence at the scale of a civil war requires significant resources. In their view, civil wars are more likely to occur in times of relative abundance. While "riots and protests, may emerge from conditions of scarcity," they argue, "sustaining a militant organization requires considerable planning and resources" (Salehyan and Hendrix 2014, p. 240). Reasons to fight might exist. For this to turn into civil war, however, people "also need the capability to do so, and environmental scarcity may limit such capability, thus undermining the resource base necessary for mobilizing armed violence" (Salehyan and Hendrix 2014, p. 240).

A related debate concerns what Adams and colleagues have claimed to be a sampling bias in studies of the connection between environment change and armed conflict (Adams et al. 2018). Levy accepts the existence of some sampling bias but rejects the view that this bias results in an overstatement of the connection between environment change and conflict. "Knowing that case selection is biased is useful, but not a reason to lower our estimate of the climate's impact on conflict" (Levy 2018, p. 441).

In responding to Levy's criticism, authors claiming bias wrote that they did not "deny a link between climate change and conflict in principal. Indeed, some of our own recent work indicates that such a link exists, but it is highly conditional." Their problem with the research being done in this field was that "sampling biases… increase the risk that such links are overstated, that crucial world regions do not receive sufficient attention and that little knowledge is produced on peaceful adaptation" (Ide et al. 2018, p. 442–3).

After reviewing the literature on the relationship between climate change and violent conflict, Sakaguchi, Varughese and Auld concluded that the "current literature offers mixed evidence. This makes it difficult to render a definitive statement about the climate-conflict relationship" (2017, p. 640). While they pointed out that just over 60% of the studies they reviewed found "that climate change variables are positively correlated with higher levels of violent conflict," Sakaguchi, Varughese and Auld also argued that "many subtleties and countertrends underlie this overall pattern" (2017, p. 640). Thus, even though "a majority of reviewed studies envision climate variables influencing conflict through a causal pathway, … these pathways are often theoretically underspecified and have only weak empirical support" (Sakaguchi et al. 2017, p. 641).

As Koubi put it, the research that has been done on this question "provides some evidence that climatic changes could act as a 'threat multiplier' in several of the world's regions. In particular, the extant literature shows that climatic conditions can lead to conflict in agricultural-dependent regions and in combination and interaction with other socioeconomic and political factors" (Koubi 2018, p. 200). After having claimed that, to their knowledge, "no one in the field of climate research has suggested that climate change could be the 'sole cause' of war, violence, unrest or migration", Butler and Kefford recommended "viewing climate change instead as a risk multiplier, influencer or co-factor … In this way of thinking, environmental and ecological factors interact with social determinants, including those that are economic, demographic and political, to produce phenomena such as migration, conflict and famine" (2018, p. 587).

There can be no doubt that conflict will increase during the final hundred years of humanity. But it will result from a complex interaction of socio-political factors and a catastrophically changed environment. It may not go beyond conflict between different groups or between the government and opposition groups and become civil war. This depends on the capacity of those opposition groups. In many cases, they will lack the resources to conduct a civil war. The Syrian war is itself a good illustration of the problem, as the groups opposed to the Syrian government have only been able to conduct the extended civil war in which they have been engaged with the support of outside groups. (Mazzetti and Apuzzo 2016).

The question of whether civil war will break out is something that can only be answered "region by region" and the answer must be based on "knowledge of pre-conflict geographies, such as drivers of resilience and vulnerability" (Farbotko 2018). Sometimes governments may abandon territory and opposition groups can seize control of that land. But it is likely to be land that is suffering worst from the effects of catastrophic human-caused environment change and will not be habitable. To replace an existing government or take control of a region within a country through civil war is no simple thing. It may happen. But it will not happen on the scale that some people have predicted. And it will not happen just because of the weather.

## International Resource Conflict

Governments in many countries will persist despite catastrophic human-caused environment change. All will be weakened and functioning under distress, however. Some will collapse, and people will have to fend for themselves or migrate. Those that do not collapse may end up competing for resources with the governments of other countries. Water will be the focus of many of these international or intergovernmental, conflicts. The exploitation of declining fish stocks will be another resource conflict and, where they occur, will be intensified due to fish migrating or changing migration patterns because of changes in water temperature. Finally, conflicts over land and resources in the Arctic and Antarctic will also be a feature of the final hundred years of humanity.

Once again, political factors will play a crucial role in whether conflict between states results from the increased stresses produced by catastrophic human-caused environment change. Political factors may produce conflict. But they may also produce compromise. Swatuk, for example, examined two cases of alleged 'water wars' in Southern Africa. One involved an intervention by the South African Defense Force in Lesotho in 1998. The other was the mobilization of Namibian and Botswanan troops along the Chobe River border because of competing claims to an island on that river. Of these, according to Swatuk, the first conflict, in which the SADF secured a dam important to South Africa's water supply, followed a pattern associated with international climate conflict. The second, which was resolved by an appeal for adjudication by the International Court of Justice, did not (Swatuk 2015, p. 220).

Feitelson and Tubi's comparison of conflicts in the Euphrates and the lower Jordan river basins between 2007 and 2010 also points to the importance of political factors. Even though the pressures in the lower Jordan River basin were greater than those in the Euphrates river basin, conflict resulted in the Euphrates basin and not the lower Jordan River basin (Feitelson and Tubi 2017, p. 44). Despite ongoing tensions between the two countries, Israel allowed increased flows of water into Jordan and Palestine via the Jordan River. On the other hand, and even though good relations and formal agreements had previously existed, Turkey and Syria did not reach agreement. The result was that Turkey allowed less water into Syria through the Euphrates than the Syrians demanded, and Syria did not allow more water to flow on to Iraq. Agricultural production declined for both countries as a result (Feitelson and Tubi 2017, p. 45).

The proximity to urban and agricultural users, which made it easier to recycle wastewater, Israel's ability to harvest water through desalination plants and the greater possibility of transferring water within the basin all conduced to the resolution of potential conflict in the lower Jordan River basin. But a crucial factor was that the Israel government had an interest in maintaining political stability in Jordan and was aware of and sensitive to international criticisms of its occupation of Palestine's West Bank and of the amount of water it was allowing into Palestine (Feitelson and Tubi 2017, p. 45).

As Feitelson and Tubi concluded, in these situations "absolute water scarcity emerges as a poor predictor of conflict" (Feitelson and Tubi 2017, p. 45). Before we leave Feitelson and Tubi's discussion, however, it is important to note that Islamic State's occupation of a central part of the Euphrates resulted, as they put it, in "the weaponization of water. That is, the dams became a focal point of struggle, and a weapon in the war. Such weaponization… reduced the Euphrates's flow even further, with significant implications for agriculture and energy production facilities downstream" (Feitelson and Tubi 2017, p. 44).

## *Water*

As Feitelson and Tubi suggested, while it may not be a simple matter of scarcity, access to freshwater will become the principal source of tension in a world affected by catastrophic human-caused environment change. Bernauer and Siegfried argued that "one of the most important social and political risks associated with climate change pertains to water availability [with] the greatest risk of international disputes and perhaps even militarized interstate conflict… likely to materialize in international water systems located in poor and politically unstable parts of the world" (2012, p. 227).

The significance of water to conflict during the final hundred years of humanity reflects both the importance of water and the number of countries that share freshwater resources. According to Levy, "the sharing of freshwater resources between countries is very common. About three-fifths of water flowing in all rivers is shared by two or more countries – in 263 rivers basins in 145 countries. Two-fifths of the world's populations live in these river basins" (Levy 2019, p. 53).

Care is required in this context, however. For, while catastrophic human-caused environment change will make the world hotter, there is no reason to assume that it will make the world drier. So, it is less a case of there being less water and more a case of changing rainfall patterns in most regions. These changing patterns will mean that droughts will persist for much longer and that severe weather events will lead to unprecedented amounts of water falling on, often parched, lands – which means that it is likely to be lost as run-off. Periods of abundant rain will cause their own problems, particularly with floods but also with greater soil erosion. Desertification and droughts, however, create the greater potential for international conflict.

The lack of water in some regions will result from the shrinking and loss of glaciers because of global warming. Changing weather patterns might produce increased rainfall, but the glaciers will not return or will not be as extensive as they once were. Thus, there will not be the quantities of snowmelt that once provided water to river systems. If these systems flow through two or more countries, governments of the country or countries upstream will face pressure to capture as much water as can be captured and to allow as little as possible to flow to downstream countries.

But whether this leads to conflict will not be determined solely by how much water is being captured by upstream countries but will also be affected by how much military force an upstream country can deploy. The principal reason that fighting did not break out when Turkey refused to allow greater flows of water into Syria and Iraq via the Euphrates, Bernauer and Siegfried argued, was "the upstream country (Turkey) [was] also the most powerful country in the system" (2012, p. 232). In contrast, while it is downstream from the other eleven countries through which the Nile River flows, Egypt's military power means that Egypt gets the water it needs.

The threat of conflict over water is most likely, in Bernauer and Siegfried's view, among countries through which the Syr Darya river flows. This is because the region is less politically stable and "the dominant economic and military powers (Uzbekistan and Kazakhstan), which also face the most severe water security risks, are located downstream, whereas the upstream country [Kyrgyzstan] is in almost total physical control of the catchment's runoff" (Bernauer and Siegfried 2012, p. 232).

Conflict over water is already occurring between China and India and will worsen as the effects of catastrophic human-caused environment change. According to Zhang, the most likely "point where water conflicts might erupt is in the Himalaya region, between the two most populous and thirsty nations in the world, China and India, which share several transboundary rivers, including the most contentious, the Brahmaputra River" (Zhang 2016, p. 155). The tensions that develop between India and China will affect the entire region. According to Xie, Zhang and Panda, China's "water interactions with neighbouring countries are regarded as having the potential to spark open and violent conflicts. The future development of Sino–Indian relations is uncertain, with questions around possible water disputes being a particular pressure point and catalyst for regional insecurity" (Xie et al. 2018, p. 2.).

The different arenas for international water conflict discussed above are the most likely sites for military confrontation during the final hundred years of humanity. But the effects of catastrophic human-caused environment change on water supply will affect other regions in which several countries share the same river system. While other, particularly political, factors will determine whether military conflict is avoided, tension around water will be increasingly impossible to contain over the course of the final hundred years of humanity.

## *Fish*

The harvesting of edible open ocean fish will be another source of international conflict during the final hundred years of humanity. As has been discussed, catastrophic human-caused environment change will cause ocean warming, sea level rise, ocean acidification, weather pattern changes and extreme weather events. This will cause some fish species to migrate, while others will decline in number (and perhaps size, though scientists are still debating this question – see, for example, Stiasny et al. (2019)).

Some fish species will adapt to changes in their environment and be less likely to migrate. But their capacity to adapt will be challenged by the extent of the changes brought on by catastrophic human-caused environment change. Donelson and colleagues recognized that large or rapid environmental change can result in greater adaptive "responses than a small or slower change…". But they also wrote that large or rapid change "may restrict or alter the plastic changes by causing a stress response that would otherwise not occur with less extreme environmental changes. Furthermore, with a relatively large environmental change, individuals may not be able to produce complete compensation of a trait with just one or two generations" (Donelson et al. 2018, p. 25). They concluded that thresholds existed for transgenerational plasticity, "which may be influenced by the magnitude of natural environmental variation in the system" (Donelson et al. 2018, p. 25).

Asch and co-authors predicted that changes in water temperature, acidity and oxygen levels in the tropical Pacific Ocean will result in the maximum potential catch falling by over 50% by the end of the twenty-first Century (Asch et al. 2018, p. 8–9). Blasiak and colleagues observed increasing conflict due to significant regional differences in vulnerability to the loss of fish stocks. The regions where the greatest variation occurs, they wrote, will "be particularly conflict-prone if exposure to the impacts of climate change on fisheries becomes more pronounced without mitigating efforts to build adaptive capacity" (Blasiak et al. 2017a, b, p. 11).

Declining fish stocks in the Persian Gulf are particularly significant, given that the region is already affected by other causes of conflict. Wabnitz and colleagues studied 55 "priority marine species" that were selected because of "their importance to fisheries and selection by local stakeholders" in the Persian Gulf. They found that habitat suitability for all 55 species would "undergo major declines" and that "local extinction rates are expected to increase considerably throughout the Gulf. Impacts are predicted to be particularly high along the south and southwestern coasts, where high rates of local extinction are projected by the end of the 21st century" (Wabnitz et al. 2018, p. 14).

Some species will not decline in numbers, though, as "climate change is expected to reduce marine primary production in some areas and increase it in others" (Brander et al. 2017, p. 49). Fish stocks in the West Africa region are likely to benefit from climate change, at least in the early years; whereas fish stocks in the South East Asian and the Pacific regions are projected to decline (Brander et al. 2017, p. 51–2). But even when there may be no overall decline, and perhaps even an increase, in the numbers of fish of a desirable species, changes to their migration patterns will produce conflict or intensify existing conflicts. Dueri and colleague's "projections for 2050 and 2095 show[ed] an increase in global skipjack biomass compared to 2010." But they also indicated "major changes in its distribution, impacting local and regional fishing efforts" (Dueri et al. 2016, p. 1).

International disputes over the size of fish catches have arisen in the past, so they will not be something new to the final hundred years of humanity. So far, a level of international cooperation to preserve fish stocks has been achieved. But some of the agreements between countries with respect to their shares of fish that move through their domestic waters are unlikely to survive catastrophic human-caused environment change.

By way of an indication of what is to come, change in the habits of the northeast Atlantic mackerel stock in 2007 resulted in conflict between the EU, Norway, the Faroe Islands, and Iceland (Spijkers and Boonstra 2017, p. 1838). A change in sea temperatures was not the sole cause of the expansion of the areas through which the fish migrated. But it was a contributing factor in the conflict over the size and allocation of fishing quotas among the countries through whose waters the mackerel migrated (Spijkers and Boonstra 2017, p. 1836).

The politics at work here meant that the governments involved disagreed with respect to explaining the expansion of the area through which the mackerel migrated.

Treating climate change as the principal cause of the expansion was in the interests of the people of Iceland and the Faroe Islands because it justified larger quotas for those countries. But accepting that the expansion resulted from climate change weakened the EU's and Norway's position, as this meant that they would have to accept a permanent reduction in their quotas (Spijkers and Boonstra 2017, p. 1847).

A permanent increase in the share of mackerel taken by Iceland, however, would give the EU and Norway an incentive to retaliate "by fishing harder, which greatly reduces the number of years when mackerel are available in the Icelandic zone" (Hannesson 2012, p. 259). So, counter-strategies were available for those countries that did not want to fully cooperate in managing the resource. An ongoing lack of cooperation between the nations affected, and cooperation will become harder to sustain as we move further into the final hundred years of humanity, will mean that everyone is likely to lose. As Hannesson warned with respect to the mackerel wars, "the total absence of cooperation could be extremely destructive, reducing the stock to near extinction" (2012, p. 268).

The North Pacific is another area in which conflict is likely to occur because of the effects of catastrophic human-caused environment change on fish migration patterns. According to Woodworth-Jefcoats, Polovina and Drazen, tuna and billfish migration will shift northward and eastward (Woodworth-Jefcoats et al. 2017, p 1003). This leaves countries in the central and western subtropical regions of the North Pacific with declining fish stocks. This general decline in tuna and billfish numbers will result from increased water temperature and a decline in volume of the plankton on which the tuna feed. So, competition and conflict will be over a potential catch that is at least 20% and up to 50% smaller than was previously available (Woodworth-Jefcoats et al. 2017, p. 1005).

The Pacific Island region will also witness a progressive eastward shift in the migration patterns of skipjack tuna. The result will be a growing gap between the fish available to feed growing human populations and the capacity to sustain harvests. The result is that shortages are expected in some Pacific nations by 2035 (Johnson et al. 2017, p. 333.).

Norwegian Spring Spawning Herring and Pacific Salmon are other examples of species whose changed behaviour will lead to international conflict. While agreements existed that were designed to ensure the sustainability of Pacific salmon, "the effects of a sustained climatic regime shift significantly altered the parties' incentives to continue cooperating. Cooperation fell apart because the allocation rules in place were not sufficiently flexible to accommodate such changes" (Miller 2007, p. 59). The story of the Norwegian Spring Spawning Herring had an additional twist, in that countries that had not participated in rebuilding the fish stocks enabled by an international agreement benefitted from a change in migration patterns. This showed that "even in relatively simple cases in which a fish stock is shared among a small number of harvesting nations, climate-driven changes in the abundance, catchability or spatial distribution of the stocks can make it difficult to maintain cooperation on effective management of harvests" (Miller 2007, p. 59).

Catastrophic human-caused environment change will create winners and losers and many of the nations that are losers will have to do something to change their situation and this will be met with resistance from the winners. It is hard to accurately predict which countries will win and which lose in all cases, though. Belhabib, Lam and Cheung, for example, cited two studies concerning West African fisheries that predicted that climate change would have opposite effects on the two major ecosystems in the region – the Guinea Current Large Marine Ecosystem and the Canary Current Large Marine Ecosystem (2016, p. 25). But when fish numbers decline in general, then the competition and conflict will become more intense.

As Brander and co-authors wrote, however, aquaculture is being promoted "as an adaptation option to offset negative impacts that climate change could have on fish catches" (Brander et al. 2017, p. 54). And certainly, aquaculture is a growing contributor to food production (particularly in Asia, which accounted for 89% of the world's aquaculture production in 2016 (Ahmed et al. 2019, p. 159). But, aquaculture "is highly dependent on the climate, and thus, climate change is a threat to increase global fish production" (Ahmed et al. 2019, p. 159). So, even if aquaculture provides early relief from pressures that might cause conflicts over fishing rights, this relief may not last.

For, as people move further into the final hundred years of humanity, the productivity of aquaculture will face increasing threats from "a combination of climatic variables, including cyclone, drought, flood, global warming, ocean acidification, rainfall variation, salinity, and sea level rise" (Ahmed et al. 2019, p. 163). Further, as Brander and colleagues wrote, "aquaculture can negatively impact on fisheries as, for example, with large-scale conversion of mangroves into fish and shrimp ponds in SE Asia that has resulted in coastal erosion and put large areas at risk from flooding. The problem is only likely to get worse as sea level rises due to climate change" (Brander et al. 2017, p. 54).

## Polar Conflict

Catastrophic human-caused environment change, principally in the form of global warming, will also result in conflict in the Arctic and Antarctic. This conflict will relate to the exploitation of resources that have been made available as polar icecaps melt in early period of the final hundred years of humanity. Later in the final hundred years of humanity the temperatures in both areas is likely to place them amongst the most livable parts of the planet. They will become places over which nations compete for control, while also becoming places to which people forced from their lands will seek to migrate. The politics of the final hundred years of humanity will involve conflict in, what some refer to as, the 'polar tension belts.'

While Cassotta and colleagues were writing about the Arctic, their comments apply equally to the Antarctic: in both cases, "fossil fuels are expected to become drivers of geopolitical changes because climate change impact facilitates vast resource extraction, increases competition between states and results in tensions and threats, even military ones" (Cassotta et al. 2016, p. 5). But conflict will not just arise with respect to oil in the Arctic, as different countries will compete to harvest fish and extract diamonds and claim shipping routes in the region. Shipping routes will not cause conflict in the Antarctic. Fish will, however, and oil and diamonds may – both are assumed to be present. Later in the final hundred years of humanity increased temperatures in the polar regions will make the lands there more conducive to human occupation and this will provide another source of international conflict as militarily powerful countries look to occupy these territories to provide food for their populations.

The differences in the claims over the Arctic and Antarctic mean that the countries involved in international conflict in those regions will be different. Five countries border the Arctic region, Russia, the U.S. (via Alaska), Canada, Norway and Denmark (via Greenland), and each will claim at least some Arctic territory. The most likely conflict will be between Russia and a coalition of the other four countries, led by the US. Their proximity and, in the case of the US and Russia, their naval and military power mean that China is the only other potential competitor likely to arise.

Given the extent of its territory that borders the Arctic region, Russia will claim the greatest share of the region. While concerns with respect to Russia's positioning itself to make claims over the Arctic region may reflect an American perspective, this perspective cannot be dismissed. First, because the only real challenger to U.S. naval power in the area comes from the Russian Northern Fleet. Given their naval power, competition over shipping lanes is most likely to be between Russia and the U.S. The second reason to believe that the Russian government is likely to cause conflicts is that it is already positioning Russia to make claims over Arctic territories. According to Holland, "Russia's military has created a Joint Strategic Command North dedicated to protecting the nation's interests in the Arctic Circle. The command has reopened cold war bases across Russia's Arctic coastline… Long-range bombers that could test American and Canadian air defenses in the Arctic are being upgraded" (Holland 2016, p. 65).

The Chinese have been late entrants to the competition over the Arctic region, but President Xi has made clear that part of his program for making China into a Great Power includes establishing rights in the Arctic and Antarctic regions. Indeed, "the goal of becoming a polar power is an important component of China's process to become a global power and is a key part of Beijing's maritime strategy" (Brady 2017, p. 3). Thus by 2017, Brady wrote, "as part of its overall expanding global foreign policy, China has become a leading polar player with wide-ranging and complex interests in both the Arctic and Antarctic" (2017, p. 2). In the case of the Arctic, the Chinese government released its first "Arctic White Paper" in January 2018. It stated that "due to its status, size and proximity to the Arctic, has legitimate interests in the region and should therefore be respected and included as an important stakeholder" (Sørensen 2018, p. 2).

Claims over the Antarctic are less a reflection of clear and direct territorial proximity and more a matter of history. Of course, New Zealand, Argentina, Chile and Australia can claim to be geographically closest to Antarctica. But the UK, France and Norway have made claims to territories near Antarctica, which allows them to claim proximity as a basis for their claims over parts of the region. In addition to the seven countries that have made claims over parts of the Antarctic region, South

Africa's claims to the Prince Edward Islands may make proximity a justification for its occupation of additional territory in the Antarctic region.

In 2008, a spokesperson for the Russian government said that it made no claims to the Antarctic region and its focus on the Arctic may mean that it does not assert interests in the Antarctic. But the Russian government, like the US government (which oversaw the signing of the Antarctic Treaty), portrays Antarctica as a resource for all people. And both Russia and the US maintain a "basis of claim" to rights in the Antarctic region under the Arctic Treaty (Secretariat of the Antarctic Treaty). The fact that the former USSR made a claim to Antarctic territory might provide another historical basis for a claim from Russia. China's push to become a Great Power (including a great Polar Power) means that it has also positioned itself to make claims over Antarctic territory by opening its fourth research base in the region in 2015, with plans to open a fifth base in 2022 (See Bender (2016) and Perlez (2015)) Even the Ukraine has made a claim to rights in the Antarctic region.

Of the seven countries with current claims to Antarctic territory, five (the UK, France, Australia, New Zealand and Norway) recognize each other's claims. Argentina and Chile's claims are not recognized because some of the territory they claim overlaps with territory claimed by the UK. Russia and the US have reserved a right to claim Antarctic territory and the Chinese will stake a claim to some territory in Antarctica (if on no other basis that they have a fifth of the world's population and should get a significant share of the resources of the Antarctic and Arctic regions).

The problem with claims to Antarctica is that only the governments of the twelve countries that signed the Antarctic Treaty consider it valid (the seven claimants to the territory plus Belgium, Japan, South Africa, the Soviet Union and the United States). The governments of other countries that do not recognize the Treaty have no reason to accept any of the claims made under it. The most important of these, in terms of military capacity, is the Chinese government. The size of its population and their vulnerability to the effects of catastrophic human-caused environment change mean that the Chinese government has a significant motive to claim parts of the Antarctic region. Though, "many of the opportunities in the Arctic and Antarctic that attract Chinese interest will not be available for several decades, and they will take considerable planning, coordination, and international diplomacy to maximize" (Brady 2017, p. 8).

An additional source of conflict will arise because, even if they recognize some territorial claims, countries that were not signatories to the Antarctic Treaty may not accept that fishing rights apply in the region in the same way that they apply to territorial waters off a nation's coast. Those governments that survive the effects catastrophic human-caused environment change will come under increasing pressure to claim parts of the Antarctic region to provide for their population. The most powerful of these will see the price paid to claim a part of the Antarctic region as one worth paying. These countries may well wait until the Antarctic Treaty expires in 2048 before they make their claims, if they can afford to wait.

The tension in polar regions will intensify during the final hundred years of humanity because of countries seeking to exploit resources and, eventually, occupy parts of these regions. The lines of conflict are somewhat clearer cut in the Arctic polar tension belt, as the countries with competing claims to the region are geographically closer than those currently claiming territory in the Antarctic region. The proximity of the countries with claims to Arctic resources means that fishing rights will be more of an extension of territorial rights than they will be in the Antarctic. Conflict in the Arctic may be more open than that in the Antarctic.

## Border Conflict

The other form of conflict that will mark the final hundred years of humanity will occur at the land and sea borders of those nations that have the capacity to protect their citizens from the worst effects of catastrophic human-caused environment change. As noted earlier, in some cases those in government in countries with a large landmass may abandon areas and narrow the territory over which they assert control. And as has also been discussed, considerable debate exists concerning whether climate change was responsible for the civil war in Syria. But whether it was a civil war or not, the conflict in Syria is important here for two reasons.

First, because it resulted in over five million people fleeing Syria to seek refuge in other countries. This provides some indication of the scale of the movement of people to be expected during the final hundred years of humanity. More importantly, and second, the growing resistance to accepting refugees on the parts of governments of countries of the Global North is a sign of what is to come during the final hundred years of humanity. The rise of right wing parties in Europe must be traced to the pressures created by an influx of Syrian refugees. Or, at least, a perception of being overrun by refugees has produced this effect (see Bock and Macdonald 2019). The construction of a migration 'crisis', through narratives of disaster or a fear of terrorism "has arguably helped underpin a rise in populist politics and fuelled a resurgent right-wing politics across Europe" (Dempsey and McDowell 2019, p. 54).

As Dempsey and McDowell pointed out, the response to Syrian refugees has revised many people's views about what was taken to be the declining significance of national borders. At least, that is, their significance with respect to the movement of people, as borders had a decreasing influence over the international flows of goods and services. In 2019, Dempsey and McDowell wrote, "in a globalizing world, where fences and border walls were once predicted to disappear, we are witnessing a world that is not universally 'borderless'" (2019, p. 153). Borders exist, but they operate only to restrict people's movement. "While the ease with which many goods and materials are able to cross international borders grows, the movement of bodies – especially those deemed 'politically undesirable' – are increasingly facing hardened borders and territorial closures" (Dempsey and McDowell 2019, p. 153–4).

This hardening of borders, Dempsey and McDowell argued, resulted from "efforts by the privileged few who veritably benefit from what globalization purports to offer to protect that privilege" (Dempsey and McDowell 2019, p. 154). One means through which borders have hardened with respect to the movement of people has been "through the employment of narratives about terrorism and smuggling as rationale for increased policing of international borders" (Dempsey and McDowell 2019, p. 154).

## *'Climate Change Refugees'*

One strategy intended to overcome resistance to climate change migration has been to define a new type of refugee, 'climate change refugees' or 'environmental refugees' (Berchin et al. 2017), who have a right to migrate to countries that have not been devastated by the effects of catastrophic human-caused environment change (countries, we should note, given the moral issues that arise in this context, that in almost all cases have played a major role in causing that environment change). Formal recognition of the category of 'climate change *refugee*' or 'environmental *refugee*' has met and will continue to meet resistance from those who preside over governments that will remain largely stable despite the effects of catastrophic human-caused environment change. As Jolly and Ahmad argued "though international refugee law appears to be the closest mechanism to offer a protective rights framework for the mass exodus of climate change-induced displacement, the effectiveness of the refugee regime is severely handicapped by the restrictive interpretation of the term 'refugee' …" (2018, p. 6).

Tens of millions of people will be forced to seek refuge because of the effects of catastrophic human-caused environment change. Islands will sink, lands will become uninhabitable and people will be forced to migrate. They will do everything they can to stay even when their islands sink or their lands becomes desert and increasingly unlivable. But migration is not just about environmental changes. Islands will sink and lands will become deserts. But it is only when political institutions collapse that recovery and survival become impossible. The story of forced migration, then, is one of a mixture of environmental and political factors resulting in people being forced to flee their lands to seek refuge in nations coping better with the effects of catastrophic human-caused environment change.

### Sinking Islands/Burning Lands

Forced migration from sinking islands and burning lands is simply a logical extension of the foreground of catastrophic human-caused environment change depicted earlier. The threat of rising sea levels is the most obvious cause of exodus because people cannot live when their lands are flooded. The points at which desertification and unlivable temperatures forces people to flee their territories are harder to determine. The discussion in this part is intended to add border conflict to an account of the politics of the final hundred years of humanity, however, and not to provide a comprehensive and detailed analysis of those who will be forced to seek refuge in other countries because of catastrophic human-caused environment change.

Some indication of the scale of the problem of forced migration was provided in a 2014 report published by the UK Ministry of Defence that predicted that desertification would force 60 million people to flee sub-Saharan Africa by 2020 (Ministry of Defence 2014). Even when sufficient rains fall and the soil is not thoroughly degraded, surface temperatures may exceed the 37 °C (98.6 °F) that human beings can tolerate if relative humidity does not rise above 50%. (Higher humidity lowers the temperature at which people can survive.) According to Mora and co-authors, in the worst of the scenarios developed by the International Panel on Climate Change "almost three-quarters of the world's human population could be exposed to deadly climatic conditions as opposed to one-half under" the IPCC's best-case scenario (Mora et al. 2017, p. 505).

Sea levels will not rise uniformly, but nowhere is the increase likely to be much below 2 m. A list of island nations whose populations live less than 2 m above the current sea level has not yet been developed. But it includes: Kiribati, whose government has bought land in Fiji for when its 100,000 people have to relocate; the Maldives, which, at an average of 1.3 m above sea level, is the world's lowest-lying nation; the Seychelles; the Solomon Islands, which are also sinking; the Federated States of Micronesia; the Republic of Palau; Tuvalu; and the Republic of the Marshall Islands.

The countries that are most vulnerable to the effects of climate change on fish stocks include small island states in the Pacific Ocean and Caribbean and countries on the Western and Eastern coasts of Africa. But seven of the ten countries most threatened by a significant decline in fish stocks are small island nations (Blasiak et al. 2017a, b, p. 7). Increasing levels of toxins and plastics will affect the fish that are caught, though, further threatening the people who consume significant quantities of marine products. According to a 2016 study, fish are drawn to eat micro-plastics in the ocean (McGrath 2016) and a later study found that these micro-plastics produce brain damage in fish (Science Daily 2017). So, fish populations will be affected by increasing water temperatures and by erratic behavior caused by brain damage.

Migration that results from desertification will be far greater than migration from sinking islands. Many countries of the Global South will be affected by the expansion of drylands and the desertification that will result from the further drying of existing drylands. According to Huang and colleagues, by the late twenty-first century "drylands will cover nearly all of the continental areas of Africa and Eurasia between 30 to 60 N and 15 to 50 S…" (Huang et al. 2016, p. 169). More importantly, 78% of the increased drylands will occur in countries of the Global South (Huang et al. 2016, p. 169). And, by the end of the twenty-first century 61% of the total area of the countries of the Global South will be drylands (Huang et al. 2016, p. 171).

Survival will not only become more difficult in those countries, catastrophic human-caused environment change will result in the collapse of many of the governments whose people are affected by desertification during the final hundred years of humanity (Huang et al. 2016, p. 169). And it is not just a problem of dry areas becoming drier and dry areas expanding in countries of the Global South. Half of the population growth that will occur during the final hundred years of humanity will occur in developing countries in drylands. Increasing populations in these developing countries will produce greater demand for productivity from drylands and these will quickly degrade, and further desertification will result (Huang et al. 2016, p. 171). As Huang and co-authors concluded: "developing countries account for more than 70% of the global land area and population, degradation and desertification will become a challenge to the global ecosystem and human survival in the near future" (Huang et al. 2016, p. 171).

## 'Climate Change Refugees'?

One of the most important political issues that have arisen in this context during the last ten years has been the status of people seeking to be treated as 'climate change refugees.' Despite a number of attempts to establish the category 'climate change refugees' in a variety of international forums and courts of law, the category has no legal status. Indeed, as Sahinkuye argued, "people whose environments are destroyed by ecological disasters remain largely ignored at the global level." Thus, even in 2019 we are still asking the question "Do people facing such disasters have a status under international law?" (Sahinkuye 2019, p. 12).

The people leaving Papua New Guinea's Carteret Islands in 2014 were claimed to be among the first, if not the first, "official climate change refugees" (Merchant 2014). There was nothing 'official' about their status, of course. 'Climate change refugee' was an invention of those claiming refuge, of their advocates and of journalists. The fact that Sahinkuye asked the question of the status of victims of environmental disasters under international law in 2019 means that this status is not recognized in international law and it has never been accepted in crucial international forums.

Another problem with attempts to claim the status of 'climate change refugee' is that climate change is not and will never be the sole factor in driving people from their homes. As Abel and co-authors argued, "climate change… will not generate asylum seeking everywhere but [is] likely in a country undergoing political transformation where conflict represents a form of population discontent towards inefficient responses of the government to climate impacts" (Abel et al. 2019, p, 246). Climate change refugees seek refuge because their governments cannot protect them from the effects of catastrophic human-caused environment change. "It is typically a lack of support mechanisms to deal with climate change impacts, rather than the climate change or its impacts per se, which forms the root causes [of forced migration]" (Kelman 2019, p. 11).

As has been noted, the category 'climate change refugee' has not been recognized in international law. But the questions concerning the rights of people forced to leave their countries because of catastrophic human-caused environment change remain important ones. The key question being whether the governments of the countries in which they are seeking refuge can be required, and perhaps forced, to respect these rights. For if the people forced to leave their countries due to catastrophic human-caused environment change have legal rights, then they may be able to enforce those rights.

An important element of the politics that arises in this context, then, relates to the naming of those forced to flee their countries because of catastrophic human-caused environment change. This is not just about word play. There is a politics to how we name those forced to flee. As Kelman argued, "labelling or not labelling people as 'climate change migrants' or variations is a subjective, and hence political, choice" (Kelman 2019, p. 12). Whether they are called 'migrants' or 'refugees' is important because it relates to the question of whether those claiming to have been forced to flee and have not chosen to leave. For whether they have, indeed, been forced to flee because of the effects of catastrophic human-caused climate change is a crucial matter.

Given that no government is persecuting 'climate change refugees' then their decision to leave may not have been forced. There is a clear and direct relation between 'climate change refugee' and 'forced migration'; but there is no like relationship between 'climate change refugee' and 'voluntary migration'. So, the question of whether those seeking to be recognized as 'climate change refugees' have been forced to seek refuge in another country becomes the focus of discussion and the lens through which those seeking to migrate are viewed. Kelman argued that energies "would be better spent contributing to the ongoing efforts to understand people's motivations for migrating or not migrating under all circumstances... rather than starting with the assumed inevitability of climate change causing migration" (2019, p. 12).

A considerable number of those who object to the use of the phrase 'climate change refugee' do not want to create legal rights for people who must abandon their countries due to the effects of catastrophic human-caused environment change. Others think that referring to people as 'climate change refugees' weakens the position of those seeking refuge because it makes them appear vulnerable and unable to control the situation in which they find themselves. Yet others argue that using the term 'refugee' might help to highlight the fact that the people seeking refuge have played little or no role in causing the problems that are forcing them to flee. To call them 'migrants', from this point of view, is to treat them as choosing to leave and not as being forced to leave because of the actions of those in countries of the Global North, who have principal responsibility for catastrophic human-caused environment change.

The UN Refugees Convention creates rights for people who are forced to escape persecution by their governments or persecution by other members of society from which their government is not protecting them. So, the problems these 'refugees' face are a result of wrongdoing by those who are relatively proximate to them. Those displaced due to catastrophic human-caused environment change are rarely the ones primarily responsible for producing greenhouse gases and toxins (though they may bear some responsibility for species depletion). As has been noted above, many of these people have been living simple, largely environmentally friendly, lives and may reject the high-pressure (and high-carbon, high-toxin and high-species depleting) lives of those in countries of the Global North. Their problems are not locally caused, and they have stayed where they are because they do not want to live in the societies that are largely responsible for catastrophic human-caused environment change.

But they are forced to seek refuge in those countries because it makes little sense for them to go to other poorer less-developed countries of the Global South, as they will place additional demands on limited resources in those countries. In most cases, then, refugees will seek refuge amongst those who have contributed more to catastrophic human-caused environment change than those seeking refuge. A generous view of the situation is that it is akin to asking to live with neighbours who burnt down your house without meaning to do so. It is generous because those in countries of the Global North were alerted to problems with respect to catastrophic human-caused environment change decades ago and have, through their governments, chosen to do little to nothing with respect to changing their ways. But it is hard to imagine wanting to move in with neighbours who deliberately or even negligently burnt down your house.

The point is that they are asking to live with their neighbours because those neighbours burnt down their house. This is the key to the claim to be a 'climate change refugee'. Those seeking refuge would not be doing if those in countries of the Global North had not done what they did and, in many instances, continue to do. And they would have chosen not to have their house burnt down had they a choice in the matter. But they did not have a choice, and, because of other people's actions, they must move.

When asked, people living in island nations threatened by catastrophic human-caused environment change indicate that they would rather stay. They do not want to lose all the things they must sacrifice when they leave their lands. One Tuvaluan has been quoted as saying that his people wanted to maintain their culture and, on that basis, that he preferred to "just stay here and die" (Constable 2017).

A crucial difference between 'climate change refugees' and 'refugees' under the 1951 UN Convention is that the governments from which refugees seek haven under the Convention are not at fault. At most, they might have engaged in the wars that forced refugees to flee from the dangerous situation that was created. The basis of the 1951 Convention, though, is a humanitarian principle that protection should be offered to those who will not find it in the country in which they live. Claiming rights as a 'climate change refugee' is different. It is more like demanding that governments in countries that survive its effects acknowledge their responsibility for causing catastrophic human-caused environment change. Other types of refugees do not demand that those from whom they seek refuge accept responsibility for the situation that has given rise to a need to seek refuge. People demanding rights as refugees because of catastrophic human caused climate change want recognition that their need to move was caused by the people who live in the countries to which they want to migrate.

Courts in New Zealand and Australia, which were the first jurisdictions in which this was tested, decided that that climate changes is not basis on which someone can apply to be declared a refugee "because a person can only claim to be a refugee under the 1951 Refugee Convention" (Constable 2017). The problem for those claiming to be 'climate change refugees' under this Convention is that it requires that someone applying to be declared a refugee must be in danger of persecution. Persecution can result from "an act of Government against individuals" and arises is situations in which a Government is unwilling to protect persecuted citizens. Importantly, the Convention restricts persecution to being with respect to "race, religion, nationality, membership of a particular social group or political opinion" (United Nations General Assembly 1951, art. 1).

Akakpo and Lenard's solution to problems of applying the UN Refugee Convention to people forced to seek refuge due to catastrophic human-caused environment change is to follow the underlying principles of the Convention. They argued for the creation of a category of 'climate change refugee' for people forced to cross an international border because they face "a threat that is likely to be of indefinite duration, and where the specific threat is to the ability to live a decent life." Akakpo and Lenard saw this as a solution because, "just as in the case of those already captured by the Geneva Convention's definition of a refugee, these climate migrants are forced to cross borders against their will" (Akakpo and Lenard 2014, p. 500). So far, though, the UN has not accepted this approach.

This does not mean that organizations like the United Nations High Commission for Refugees are ignoring the issue of climate change-related migration as a future policy issue. They simply recommend that this is an issue of migration and that potential climate change migrants are not referred to as 'refugees' (Constable 2017, p. 1034). Their objective is to divorce the category of 'migrant' from that of 'refugee' in order to reserve the use of 'refugees' for the people covered by the 1951 Refugee Convention.

But it is not just those in organizations like the UNHCR who want to restrict the term 'refugee' to people forced to seek refuge in another country because of persecution. Many of those who seek to migrate because of catastrophic human-caused environment change do not want to be called refugees. They do not see themselves as powerless, as victims of persecution are seen, and do not want to be treated as if they have no agency and lack power over their own lives (Hingley 2017, p. 164). In contrast, it is useful to appear weak and vulnerable when applying as 'refugees' under the 1951 UN Refugee Convention. To do so, as Wellman and Cole suggested, makes a refugee's situation appear morally equivalent to "that of a baby who has been left on ones doorstep in the dead of winter. Only a moral monster would deny the duty to bring this infant into her home, and no theorist who endorses human rights could deny that states must admit refugees" (2011, p. 120).

While the question of how many people will be forced to leave their countries because of the effects of catastrophic human-caused environment change remains open and is likely hard to resolve (see Kelman (2019)), questions about how to deal with these people are not. The possibilities range from making a deliberate decision not to recognize any right to migrate to another country due to catastrophic human-caused environment change to creating a legal right to be taken in by the government of another country for those properly declared 'climate change refugee'.

Resistance to recognizing rights to refuge for people forced to flee their countries because of catastrophic human-caused environment change has been evident from the very beginning of discussions concerning international responsibility for people displaced due to environmental factors. While the rights of those displaced by civil war or who leave their country due to persecution were recognized in the UN's Convention Relating to the Status of Refugees in 1951, the leaders of national governments have resisted recognizing climate change refugees.

During a 60th anniversary commemoration of the 1951 Refugee Convention UN officials offered to initiate a process for developing a similar convention with respect to people who might be designated refugees due to climate change and natural disasters. This offer was rejected and the governments of only five countries (Norway, Switzerland, Costa Rica, Germany and Mexico) willingly supported the UN's work on this project. Norway and Switzerland then moved to create the Steering Group of the Nansen Initiative (McAdam 2016). This Steering Group includes Norway, Sweden, Australia, Bangladesh, Costa Rica, Germany, Kenya, Mexico and the Philippines.

The goal of those involved in the Nansen Initiative is not to extend the UN Refugee Convention to create a set of rights that people forced to migrate because of catastrophic human-caused environment change can pursue through courts. The countries involved reject the application of the term 'climate refugee' to people forced to migrate because of desertification and rising sea levels. One ground for their rejection was that the concept has no place in international law. But the other ground is that the aim of the countries involved in the Nansen Initiative is the limited one of encouraging "consensus among States on key principles and elements to protect people displaced across borders in the context of disasters caused by natural hazards, including those linked to climate change" (Secretariat Nansen Initiative).

Despite various attempts to derive a right to migrate from existing legal principles (e.g., Stephens 2014), no such right exists in international law. And principles of state sovereignty make creating such a right politically impossible. A right to demand refuge exists if someone is declared to be a refugee. But this right has not been extended to people claiming to be 'climate change refugees.' The countries involved in the Nansen Initiative are working toward a middle ground. This involves creating a moral principle that limits a country's right to simply refuse entry to people forced to leave their countries because of catastrophic human-caused environment change but does not create any legal rights for those people.

## Justifying Exclusion

This brings us to the issue of justified exclusion. Attempts to develop legal or moral principles that would require governments to accept people forced to leave their homelands because of catastrophic human-caused environment change will meet resistance. Those in positions of political power will contest any process that would put pressure on them to accept these people. The most effective way to resist is to assert a right to exclude people. "Critically, because the nation-state holds exclusive control over the bodies in its territory, the right of asylum following from international law is generally understood as a right for the state to grant or deny, rather than the right of an individual to claim" (Helsin et al. 2019, p. 245).

The first step in justifying exclusion is to negate the moral force of the claims of those forced to flee because of catastrophic human-caused environment change and to shore up the force of the principle of national sovereignty as the basis for justified exclusion. Ensuring that a right to migrate is not enshrined in international law is a second stage of the process. The final stage is to assert a right on the parts of those in government to exclude people they do not wish to allow into their countries. The 1951 UN Refugee Convention has made it impossible for those in government to assert national sovereignty as the basis for refusing the claims of people who meet the criteria established under the Convention. This is why ensuring that these criteria are not extended to create a category of 'climate change refugees' who might acquire legal rights under the 1951 Convention has been an important part of justifying the exclusion of people who claim to be refugees of this type.

A significant problem arises in situations in which civil war forces people to flee. In these circumstances those seeking refuge can do so under the 1951 Convention. This makes withdrawing ratification of the Refugee Convention the only means for refusing refuge to people forced to leave their countries because of civil war that has resulted from conditions that have deteriorated because of catastrophic human-caused environment change. One of the bases for withdrawing from the UN Refugee Convention is that remaining in it means giving up sovereignty, in the sense of the people's and government's right to exclude any people they choose to exclude (rather than as a protection from interference from other states). The principal benefit of establishing a right to exclude is that it will reduce any moral pressure to accept refugees that remains despite assertions of national sovereignty.

In a book published in 2011, Wellman and Cole outline a relatively simple process for establishing a right to exclude on the parts of those who govern liberal-democracies. The first step is to reassert national sovereignty, which has been somewhat undermined by recent processes of globalization. Wellman and Cole suggested that those in government can do this by arguing that "legitimate states are entitled to political self-determination…" (2011, p. 13). The next step is to assert that people in a sovereign nation have the right to choose the people with whom they mix. In Wellman and Cole's formulation, this is to declare that "freedom of association is an integral component of self-determination…" (2011, p. 13). The final step is to claim that the right to choose with whom we mix is also the right to choose those with whom we do not mix. This, for Wellman and Cole, is to assert that "freedom of association entitles one to not associate with others" (Wellman and Cole 2011, p. 13). The result, if we follow this line of reasoning is "that legitimate states may choose not to associate with foreigners, including potential immigrants, as they see fit" (Wellman and Cole 2011, p. 13).

Blake offered an interesting variation of the line of argument that Wellman and Cole developed. For Blake "the best version of the right to exclude we might develop is one that, at its heart, relies upon the moral right to be free from unwanted relationships of obligation – where we have no independent obligation to become so obligated" (Blake 2014, p. 535). This position is of interest because it addresses the obligation that some feel is owed to people who seek refuge because of the effects of catastrophic human-caused environment change. An obligation to allow others into ones country can be readily refused under other circumstances, such as when migrants seek a better life. Refusing entry to them does not involve dismissing their human rights. Blake himself acknowledged that governments "must respect the rights of all persons to have their human rights adequately protected. States may be allowed to exclude, but they cannot exclude very much" (Blake 2014, p. 535).

Even if it was adopted, Blake's approach is unlikely to persist under conditions created by catastrophic human-caused environment change and some version of Wellman and Cole's approach is the more likely response from most governments. The primary obligation of those in positions of political power is to serve their people and the consumption of resources required for responding to the effects of catastrophic human-caused environment change will leave many unable to address the needs of those who gather on or just inside their borders. These governments are vulnerable to losing support from their people if they are thought to be neglecting their obligations to their people and not exercising their right to exclude.

## *Migrants/Refugees on the Borders*

Words are important in politics. Here they are important because they legitimize political decisions to exclude people seeking refuge from the effects of catastrophic human-caused environment change. But words cannot prevent the effects of catastrophic human-caused environment change. And increasing numbers of people will be forced to flee their homelands as those effects worsen. At some point during the final hundred years of humanity, tens of millions will have no choice but to seek refuge from those governments that have maintained stability and continue to provide for their people.

Some governments might accept climate change refugees in the early in the final hundred years of humanity. Others will not do so from the beginning. More governments will begin to refuse refuge when their resources diminish. Even if those in government are sympathetic to the needs of the people seeking refuge in their countries and have the resources necessary to support them, popular pressure may require that they refuse entry to these people. The growth of populist parties in the early decades of the twenty-first century is a result of their policies of restricting immigration. It is easy to assume that these policies will become ever more attractive to governments whose people become increasingly insecure due to the effects of catastrophic human-caused environment change.

If refugees come on boats, then the first response will be to force those boats back into international waters. If they come on land, then the initial strategy will be what Iverson called "warehousing" (Iverson 2016, p. 225). This is a strategy in which refugees are kept in camps on or just within a nation's boundaries for years, in some cases decades, awaiting admission to the countries from which they seek refuge.

Heslin and colleagues argued that "refugee camps represent the political response to mass displacement…" (Heslin et al. 2019, p. 244). They are established when the rate at which displaced populations flow into neighbouring countries exceeds "the economic, political, or social capacity or willingness of the receiving country to accommodate". But they arise "out of the juxtaposition of the international pressure to act charitably towards those in need and a state's inclination to keep 'space and distance' from the refugees themselves" (Heslin et al. 2019, p. 244). Such camps are not created as a means for assimilating those seeking refugee into the host country. Instead, the refugee camp "serves to exclude refugees' participation in the economic and social activities of the host country, as well as from political representation and participation" (Heslin et al. 2019, p. 244–5). And not only do camps play no role in determining the 'aid' those in the camps are provided, they also inhibit "the development of refugee communities into societal structures of the host country and of the global economy at large" (Heslin et al. 2019, p. 244–5).

The longer refugees are kept in camps the clearer it will become that they will never be accepted into the countries in which they seek refuge. Refugee camps are dangerous and difficult places and they become increasingly dangerous and difficult the longer people are kept in them. In many cases, catastrophic human-caused environment change will make conditions even worse. Eventually the people held in the camps will become so desperate that they will seek to force their way out of the camps and into the countries from which they seek refuge.

Of the various forms of conflict that characterize the final hundred years of humanity, the conflict on the borders at which climate change migrants or refugees gather will be the most common. The people who have been forced to leave their homelands because of the effects of catastrophic human-caused environment change will have few resources with which to fight. But fight they must. And governments will bring the resources necessary to keep them at bay.

All governments will seek to restrict the availability of information about the conditions of those seeking refuge and the steps being taken to keep them out. Public anxiety over their government's capacity to address their needs will lead most members of the public to accept the measures taken to prevent refugees from getting beyond their national borders. But they will have a strong motivation to suppress information about the deaths and injuries inflicted in keeping people from crossing their borders. Governments that are not constrained by the principles of liberal-democracy will have an easier time of it.

## Conclusion

This part began with the softer side of international conflict during the final hundred years of humanity: Political contest over the meaning of 'refugee' and the rights of migrants give over to assertions of a right to exclude. These conflicts over words are important. But they are not like the border conflicts that will become increasingly common during the final hundred years of humanity. The consequences of these border conflicts will be morally and politically challenging. This will be so even when governments have prevented the recognition of climate change refugees, the acceptance of rights associated with climate change migration and the establishment of a right to exclude. The point of each of these strategies is to refuse moral responsibility for addressing the needs of people who have been forced from their homelands because of catastrophic human-caused environment change.

While other write of 'climate wars,' the concept is of limited validity, except for conflict between nations over resources – including conflict over the resources available in Polar Regions. But only powerful nations will be able to engage in these wars. Civil wars will occur and will be associated with the effects of catastrophic human-caused environment change. They are as much a result of political instability as they are a direct result of catastrophic human-caused environment change, however. And, in addition, only well-resourced rebels will be able to vie with their governments for control of a country or region. Border conflict, as suggested, will be both the most likely and most prevalent form of conflict during the final hundred years of humanity. In some cases, it will amount to the genocide of a people seeking to escape the effects of catastrophic human-caused environment change. This will truly mark the end of humanity.

## References

Abel, G. J., Brottrager, M., Cuaresma, J. C., & Muttarak, R. (2019). Climate, conflict and forced migration. *Global Environmental Change*, *54*, 239–249. https://doi.org/10.1016/j.gloenvcha.2018.12.003.

Adams, C., Ide, T., Barnett, J., & Detges, A. (2018). Sampling bias in climate–conflict research. *Nature Climate Change*, *8*(3), 200. https://doi.org/10.1038/s41558-018-0068-2.

Ahmed, N., Thompson, S., & Glaser, M. (2019). Global aquaculture productivity, environmental sustainability, and climate change adaptability. *Environmental Management, 63*(2), 159–172. https://doi.org/10.1016/j.scitotenv.2018.10.163.

Akakpo, C. E. G., & Lenard, P. T. (2014). New challenges in immigration theory: An overview. *Critical Review of International Social and Political Philosophy, 17*(5), 493–502, https://doi.org/10.1080/13698230.2014.919055.

Asch, R. G., Cheung, W. W., & Reygondeau, G. (2018). Future marine ecosystem drivers, biodiversity, and fisheries maximum catch potential in Pacific Island countries and territories under climate change. *Marine Policy*, *88*, 285–294. https://doi.org/10.1016/j.marpol.2017.08.015.

Belhabib, D., Lam, V. W., & Cheung, W. W. (2016). Overview of West African fisheries under climate change: Impacts, vulnerabilities and adaptive responses of the artisanal and industrial sectors. *Marine Policy*, *71*, 15–28. https://doi.org/10.1016/j.marpol.2016.05.009.

Bender, J. (2016, January 6). Russia and China are drastically scaling up their presence in Antarctica, *Business Insider* https://www.businessinsider.com.au/russia-china-scaling-up-antarctic-presence-2015-12?r=US&IR=T Accessed 22 June 2019.

Berchin, I. I., Valduga, I. B., Garcia, J., & de Andrade, J. B. S. O. (2017). Climate change and forced migrations: An effort towards recognizing climate refugees. *Geoforum*, *84*, 147–150. https://doi.org/10.1016/j.geoforum.2017.06.022.

Bernauer, T., & Siegfried, T. (2012). Climate change and international water conflict in Central Asia. *Journal of Peace Research*, *49*(1), 227–239. https://doi.org/10.1177/0022343311425843.

Blake, M. (2014). The right to exclude. *Critical Review of International Social and Political Philosophy*, 17(5), 521–537, https://doi.org/10.1080/13698230.2014.919056.

Blasiak, R., Spijkers, J., Tokunaga, K., Pittman, J., Yagi, N., & Österblom, H. (2017a). Climate change and marine fisheries: Least developed countries top global index of vulnerability. *PloS one*, *12*(6), e0179632. https://doi.org/10.1371/journal.pone.0179632.

Blasiak, R., Spijkers, J., Tokunaga, K., Pittman, J., Yagi, N., & Österblom, H. (2017b). Climate change and marine fisheries: Least developed countries top global index of vulnerability. *PloS one*, *12*(6), e0179632. https://doi.org/10.1371/journal.pone.0179632.

Bock, J. J., & Macdonald, S. (Eds.). (2019). *Refugees Welcome?: Difference and Diversity in a Changing Germany*. Berghahn Books.

Brady, A. M. (2017). *China as a polar great power*. Cambridge University Press.

Brander, K., Cochrane, K., Barange, M., & Soto, D. (2017). Climate change implications for fisheries and aquaculture. In B.F. Phillips & M. Pérez-Ramírez. *Climate Change Impacts on Fisheries and Aquaculture: A Global Analysis*, *1* (pp. 45–62). Hoboken: John Wiley & Sons. https://doi.org/10.1002/9781119154051.ch23.

Buhaug, H. (2016). Climate change and conflict: taking stock. *Peace Economics, Peace Science and Public Policy*, *22*(4), 331–338. https://doi.org/10.1515/peps-2016-0034.

Butler, C. D., & Kefford, B. J. (2018). Climate change as a contributor to human conflict. *Nature*, *555*(7698), 587. https://doi.org/10.1038/d41586-018-03795-0.

Cassotta, S., Hossain, K., Ren, J., & Goodsite, M. E. (2016). Climate Change and Human Security in a Regulatory Multilevel and Multidisciplinary Dimension: The Case of the Arctic Environmental Ocean. In W. L. Filho, H. Musa, G. Cavan, P. O'Hare & J. Seixas (Eds.), *Climate Change Adaptation, Resilience and Hazards* (pp. 71–91). Cham: Springer. https://doi.org/10.1007/978-3-319-39880-8_5.

Conca, K. (2019). Is There a Role for the UN Security Council on Climate Change?. *Environment: Science and Policy for Sustainable Development*, *61*(1), 4–15. https://doi.org/10.1080/00139157.2019.1540811

Constable, A. L. (2017). Climate change and migration in the Pacific: options for Tuvalu and the Marshall Islands. *Regional environmental change*, *17*(4), 1029–1038. https://doi.org/10.1007/s10113-016-1004-5.

Dempsey, K. E., & McDowell, S. (2019). Disaster depictions and geopolitical representations in Europe's migration 'Crisis'. *Geoforum*, *98*, 153–160. https://doi.org/10.1016/j.geoforum.2018.11.008.

Donelson, J. M., Salinas, S., Munday, P. L., & Shama, L. N. (2018). Transgenerational plasticity and climate change experiments: Where do we go from here?. *Global Change Biology*, *24*(1), 13–34. https://doi.org/10.1111/gcb.13903.

Dueri, S., Guillotreau, P., Jiménez-Toribio, R., Oliveros-Ramos, R., Bopp, L., & Maury, O. (2016). Food security or economic profitability? Projecting the effects of climate and socioeconomic changes on global skipjack tuna fisheries under three management strategies. *Global environmental change*, *41*, 1–12. https://doi.org/10.1016/j.gloenvcha.2016.08.003.

Dyer, G., (2008), *Climate Wars*. Canada: Random House.

Farbotko, C. (2018). Climate change and national security: an agenda for geography. *Australian Geographer*, *49*(2), 247–253. https://doi.org/10.1080/00049182.2017.1385119.

Feitelson, E., & Tubi, A. (2017). A main driver or an intermediate variable? Climate change, water and security in the Middle East. *Global environmental change*, *44*, 39–48. https://doi.org/10.1016/j.gloenvcha.2017.03.001.

Froese, R., & Schilling, J. (2019). The Nexus of Climate Change, Land Use, and Conflicts. *Current Climate Change Reports*, *5*(1), 24–35. https://doi.org/10.1007/s40641-019-00122-1.

Gleick, P. H. (2017). Climate, water, and conflict: Commentary on Selby et al. 2017. *Political Geography*, *60*, 248–250. https://doi.org/10.1016/j.polgeo.2017.06.009.

Hannesson, R. (2012). Sharing the northeast Atlantic mackerel. ICES Journal of Marine Science, 70(2), 259–269. https://doi.org/10.1093/icesjms/fss134.

Heslin, A., Deckard, N. D., Oakes, R., & Montero-Colbert, A. (2019). Displacement and resettlement: understanding the role of climate change in contemporary migration. In R. Mechler, L.M. Bouwer, T. Schinko, S. Surminski & J-A. Linnerooth-Bayer. *Loss and Damage from Climate Change* (pp. 237–258). Springer, Cham.

Hingley, R. (2017). 'Climate Refugees': An Oceanic Perspective. *Asia & the Pacific Policy Studies*, *4*(1), 158–165. https://doi.org/10.1002/app5.163.

Holland, A. (2016). Preventing Tomorrow's Climate Wars. *Scientific American, 314*(6), 60–65. https://doi.org/10.1038/scientificamerican0616-60.

https://www.nytimes.com/2015/05/04/world/asia/china-pursuing-strategic-interests-builds-presence-in-antarctica.html. Accessed 22 June 2019.

Huang, J., Yu, H., Guan, X., Wang, G., & Guo, R. (2016). Accelerated dryland expansion under climate change. *Nature Climate Change, 6*(2), 166. https://doi.org/10.1038/NCLIMATE2837.

Ide, T., Barnett, J., Detges, A., & Adams, C. (2018). Reply to 'Sampling bias does not exaggerate climate–conflict claims'. *Nature Climate Change, 8*(6), 442. https://doi.org/10.1038/s41558-018-0171-4.

Iverson, E. L. (2016). Permanently Waiting: The Kenyan State and the Refugee Protection Regime. In N. Stokes-DuPass & R. Fruja. (Eds.), *Citizenship, Belonging, and Nation-States in the Twenty-First Century* (pp. 221–246). Palgrave Macmillan, New York.

Johnson, J. E., Bell, J. D., Allain, V., Hanich, Q., Lehodey, P., Moore, B. R., … & Senina, I. (2017). In B.F. Phillips & M. Pérez-Ramírez. *Climate Change Impacts on Fisheries and Aquaculture: A Global Analysis, 1* (pp. 333–379). Hoboken: John Wiley & Sons. https://doi.org/10.1002/9781119154051.ch11

Jolly, S., & Ahmad, N. (2018). *Climate refugees in South Asia: Protection under international legal standards and state practices in South Asia.* Springer.

Kelman, I. (2019). Imaginary Numbers of Climate Change Migrants? *Social Sciences, 8*(5), 131. https://doi.org/10.3390/socsci8050131.

Klöck, C., & Nunn, P. D. (2019). Adaptation to Climate Change in Small Island Developing States: A Systematic Literature Review of Academic Research. *The Journal of Environment & Development*, 1070496519835895. https://doi.org/10.1177/1070496519835895.

Koubi, V. (2018). Exploring the relationship between climate change and violent conflict. *Chinese Journal of Population Resources and Environment, 16*(3), 197–202. https://doi.org/10.1080/10042857.2018.1460957.

Levy, B. S. (2019). Water and Armed Conflict. In J. M. Selendy (Ed.), *Water and sanitation-related diseases and the environment: challenges, interventions, and preventive measures* (pp. 53–58). Hoboken: John Wiley & Sons.

Levy, M. (2018) Sampling bias does not exaggerate climate–conflict claims *Nature Climate Change, 8*(6), 442. https://doi.org/10.1038/s41558-018-0170-5.

Lund University. (2017 September 25). Brain damage in fish from plastic nanoparticles in water. *ScienceDaily*. www.sciencedaily.com/releases/2017/09/170925104730.htm. Accessed 22 June 2019.

Mann, M. E. (2017). Climate change: Al Gore gets inconvenient again. Nature, 547(7664), 400. https://doi.org/10.1038/547400a.

Mazzetti, M., & Apuzzo, M. (2016, January 23). US relies heavily on Saudi money to support Syrian rebels. *New York Times*. http://www.nytimes.com/2016/01/24/world/middleeast/us-relies-heavily-on-saudi-money-to-support-syrian-rebels.html?ref=topics&_r=0. Accessed 22 June 2019.

McAdam, J. (2016). From the Nansen initiative to the platform on disaster displacement: Shaping international approaches to climate change, disasters and displacement. *UNSWLJ, 39*, 1518. www.austlii.edu.au/au/journals/UNSWLJ/2016/58.html. Accessed 22 June 2019.

McGrath, M. (2016, June 2). Fish eat plastic like teens eat fast food, researchers say. *Science Magazine*. http://science.sciencemag.org/content/352/6290/1172.

Merchant, B. (2014, April 6). First official climate change refugees evacuate their island homes for good. *Earth First Journal*. http://earthfirstjournal.org/newswire/2014/04/06/first-official-climate-change-refugees-evacuate-their-island-homes-for-good/. Accessed 22 June 2019.

Miles-Novelo, A., & Anderson, C. A. (2019). Climate Change and Psychology: Effects of Rapid Global Warming on Violence and Aggression. *Current Climate Change Reports, 5*(1), 36–46. https://doi.org/10.1007/s40641-019-00121-2.

Miller, K. A. (2007). Climate variability and tropical tuna: Management challenges for highly migratory fish stocks. *Marine Policy, 31*(1), 56–70. https://doi.org/10.1016/j.marpol.2006.05.006.

Ministry of Defence, Strategic Trends Programme. 2014. Global Strategic Trends – Out to 2045. Fifth Edition. http://www.ieee.es/Galerias/fichero/OtrasPublicaciones/Internacional/2014/Global_Strategic_Trends_-_Out_to_2045.pdf. Accessed 22 June 2019.

Mora, C., Dousset, B., Caldwell, I. R., Powell, F. E., Geronimo, R. C., Bielecki, C. R., … & Lucas, M. P. (2017). Global risk of deadly heat. *Nature Climate Change, 7*(7), 501–507. https://doi.org/10.1038/nclimate3322.

Olsson, O. (2016). Climate change and market collapse: A model applied to Darfur. *Games*, 7(1), 9. https://doi.org/10.3390/g7010009.

Perlez, J. (2015 3 May). China, Pursuing Strategic Interests, Builds Presence in Antarctica. *New York Times*.

Raleigh, C., Choi, H. J., & Kniveton, D. (2015). The devil is in the details: An investigation of the relationships between conflict, food price and climate across Africa. *Global Environmental Change*, *32*, 187–199. https://doi.org/10.1016/j.gloenvcha.2015.03.005.

Sahinkuye, M. (2019). A Theoretical Framework for the Protection of Environmental Refugees in International Law. *The Transnational Human Rights Review*, *6*(1), 1. https://digitalcommons.osgoode.yorku.ca/thr/vol6/iss1/1. .

Sakaguchi, K., Varughese, A., & Auld, G. (2017). Climate wars? A systematic review of empirical analyses on the links between climate change and violent conflict. *International Studies Review*, *19*(4), 622–645. https://doi.org/10.1093/isr/vix022.

Salehyan, I., & Hendrix, C. S. (2014). Climate shocks and political violence. *Global Environmental Change*, *28*, 239–250. https://doi.org/10.1016/j.gloenvcha.2014.07.007.

Schwartz, P., & Randall, D. (2003). *An abrupt climate change scenario and its implications for United States national security*. CALIFORNIA INST OF TECH PASADENA JET PROPULSION LAB. https://apps.dtic.mil/dtic/tr/fulltext/u2/a469325.pdf. Accessed 22 June 2019.

Secretariat Nansen Initiative, https://www.nanseninitiative.org/secretariat/. Accessed 22 June 2019.

Secretariat of the Antarctic Treaty, The Antarctic Treaty http://www.ats.aq/e/ats.htm. Accessed 22 June 2019.

Selby, J. (2018). Climate change and the Syrian civil war, part II: the Jazira's agrarian crisis. *Geoforum*. https://doi.org/10.1016/j.geoforum.2018.06.010.

Selby, J., & Hoffmann, C. (2014). Beyond scarcity: rethinking water, climate change and conflict in the Sudans. *Global Environmental Change*, *29*, 360–370. https://doi.org/10.1016/j.gloenvcha.2014.01.008.

Selby, J., Dahi, O., Fröhlich, C., & Hulme, M. (2017). Climate change and the Syrian civil war revisited: a rejoinder. *Political Geography*, *60*, 253–255. https://doi.org/10.1016/j.polgeo.2017.08.001.

Shaffer, L. J. (2017). An anthropological perspective on the climate change and violence relationship. *Current Climate Change Reports*, *3*(4), 222–232. https://doi.org/10.1007/s40641-017-0076-8.

Sørensen, C. T. (2018). China as an Arctic Great Power. *FAK Policy Brief, January.* http://www.fak.dk/publikationer/Documents/Policy%20Brief%202018%2001%20februar%20UK.pdf. Accessed 22 June 2019.

Spijkers, J., & Boonstra, W. J. (2017). Environmental change and social conflict: the northeast Atlantic mackerel dispute. *Regional Environmental Change*, *17*(6), 1835–1851. https://doi.org/10.1007/s10113-017-1150-4.

Stephens, D. (2014). Establishing a positive right to migrate as a solution to food scarcity. *Emory Int'l L. Rev.*, *29*, 179. http://law.emory.edu/eilr/content/volume-29/issue-1/comments/positive-right-to-migrate.html.

Stiasny, M. H., Sswat, M., Mittermayer, F. H., Falk-Petersen, I. B., Schnell, N. K., Puvanendran, V., … & Clemmesen, C. (2019). Divergent responses of Atlantic cod to ocean acidification and food limitation. *Global change biology*, *25*(3), 839–849. https://doi.org/10.1111/gcb.14554.

Swatuk, L. A. (2015). Water conflict and cooperation in Southern Africa. *Wiley Interdisciplinary Reviews: Water, 2*(3), 215–230. https://doi.org/10.1002/wat2.1070.

Tertrais, B. (2011). The climate wars myth. The Washington Quarterly, 34(3), 17–29. https://doi.org/10.1080/0163660X.2011.587951

United Nations General Assembly (1951). *Convention and protocol relating to the status of refugees*. Art. 1. https://www.unhcr.org/3b66c2aa10.

Wabnitz, C. C., Lam, V. W., Reygondeau, G., Teh, L. C., Al-Abdulrazzak, D., Khalfallah, M., … & Cheung, W. W. (2018). Climate change impacts on marine biodiversity, fisheries and society in the Arabian Gulf. *PloS one*, *13*(5), e0194537. https://doi.org/10.1371/journal.pone.0194537.

Wellman, C. H., & Cole, P. (2011). *Debating the ethics of immigration: Is there a right to exclude?* Oxford University Press. http://eprints.uwe.ac.uk/id/eprint/21860.

Welzer, H. (2012), *Climate Wars*. Cambridge: Polity Press.

Wodak, R. (2019). Entering the 'post-shame era': the rise of illiberal democracy, populism and neo-authoritarianism in Europe. *Global Discourse: An interdisciplinary journal of current affairs*, *9*(1), 195–213. https://doi.org/10.1332/204378919X15470487645420.

Woodworth-Jefcoats, P. A., Polovina, J. J., & Drazen, J. C. (2017). Climate change is projected to reduce carrying capacity and redistribute species richness in North Pacific pelagic marine ecosystems. *Global change biology*, *23*(3), 1000–1008. https://doi.org/10.1111/gcb.13471.

Xie, L., Zhang, Y., & Panda, J. P. (2018). Mismatched Diplomacy: China–India Water Relations Over the Ganges–Brahmaputra–Meghna River Basin. *Journal of Contemporary China*, *27*(109), 32–46. https://doi.org/10.1080/10670564.2017.1363014.

Zhang, H. (2016). Sino-Indian water disputes: the coming water wars?. *Wiley Interdisciplinary Reviews: Water*, *3*(2), 155–166. https://doi.org/10.1002/wat2.1123.

# Chapter 7
# The Politics of Desperate Measures

At a certain point in its progression, the effects of catastrophic human-caused environment change will become so severe that climate change denial will no longer be effective, most resistance to policies to address catastrophic human-caused environment change will have dissipated and people will turn to their governments and demand action. Meaningful action on the parts of those in government will be enabled at this point because catastrophic human-caused environment change threatens the status quo. And, as was explained in Chap. 2, the principal, perhaps sole, purpose of political institutions is to defend the status quo, which is the defense of socially dominant interests.

The simplest immediate option for those in government is to introduce policies designed to achieve radical population reduction. While they may have little effect, they symbolize action on the parts of those in government. The policies that could be introduced to attain this end include: mandatory contraception, readily available abortion, voluntary and involuntary euthanasia and a one-child policy (with forced sterilization constituting a point at or beyond the limits of the acceptable). Policies to radically reduce the population are not the only policies likely to be introduced during the final hundred years of humanity, of course. Other policies that may be implemented include the introduction of and/or increased use of nuclear energy and the promotion of the widespread use of heat, drought and toxin tolerant genetically modified organisms. Both policies have met with considerable opposition in the past and they remain contested. They are also of dubious utility, given the time it takes to build nuclear power plants and to introduce GMOs.

With respect to the latter, it is worth noting Tylecote's 2019 lament concerning the problems with introducing plants engineered to resist the effects of catastrophic human-caused environment change. Few GMOs were in use in 2019, despite, in Tylecote's view, "the maturing of biotechnology with genomics, proteomics and genome editing, to the point where it can take its place alongside ICT as a new techno-economic paradigm." Part of the problem, in his view, was that "plant breeding… needs a change in the stance, or a reduction in the influence, of powerful environmental groups" (Tylecote 2019, p. 866). Tylecote described a "dialectical


I. Cook, *The Politics of the Final Hundred Years of Humanity (2030-2130)*,
https://doi.org/10.1007/978-981-15-1259-9_7

perfect storm" in which a neoliberal thesis, "the profit motive must rule; almost anything must be patentable with a view to profit", met an environmentalist antithesis, "we do not trust the products of profit-seeking capitalists who want to own new life-forms." In the resulting synthesis, "profit from the safer new life-forms is reserved to the strongest capitalists, who continue to serve as useful demons for environmentalist fund-raising" (Tylecote 2019, p. 866). The introduction of GMOs is likely to be slow, then, and to be too late.

The other principal feature of the politics of the final hundred years of humanity will be the geoengineering projects that those in government and outside government will undertake in response to public demands to address the effects of catastrophic human-caused environment change. Geoengineering comes in two principal forms: negative emissions technologies (NETS) and solar radiation management (SRM). While the former is expensive, some forms of the latter are not. SRM can be as relatively simple and cheap as injecting sulfate aerosols into the stratosphere, spraying sea water into the atmosphere to whiten clouds and enhance their reflectivity or sending millions of mirrors into space to reflect solar radiation away from the earth's surface. More elaborate forms of SRM have been proposed, of which solar sails is the most elegant.

While the effectiveness of these relative simple and cheaper forms of SRM geoengineering is unclear, any government, firm or person with the necessary resources can undertake them. Some level of oversight to ensure the coordination and regulation of any geoengineering projects would be required to ensure that a collective interest was served, however, and that it continued to be served. One of the features of injecting sulfate aerosols into the stratosphere (stratospheric aerosol injection (SAI)), for example, is that this must be maintained for as long as a significant greenhouse effect would trap the radiation in the earth's atmosphere. Any sudden diminution of the level of aerosols in the stratosphere would lead to a rapid increase in the average global surface temperature.

More importantly, some SRM initiatives, including SAI, will not have a consistent effect across the entire planet – with rainfall patterns potentially being changed to the detriment of the people of some countries. Adopting these measures will require some provision for the people negatively affected by SAI initiatives. Some of the most interesting questions that arise with respect to the politics of the final hundred years of humanity relate to the governance of SAI projects. Assuming, that is, that any level of cooperation can be achieved as humanity moves further into its final hundred years.

This chapter has two parts. The first addresses the politics of radical population reduction. The second part addresses the politics of geoengineering. Geoengineering initiatives are only a means for slowing global warming. None address ocean acidification, increasing toxicity and species depletion and extinction; so, geoengineering cannot be taken to represent a means for ensuring humanity's survival. Indeed, none of the policies that will be introduced to address the effects of catastrophic human-caused environment change are likely avert a trajectory toward and into the final hundred years of humanity. The threat to the status quo, however, requires that those in government act and are seen to act.

Discussions that might be thought to be missing from this chapter include those associated with the politics of what Manes called "green rage" (1990) and Fletcher "green violence" (2018) and eco-anarchist practice. The chapter is also silent with respect to possibilities for escape into artificial terrestrial environments, located in space or on another planet (terrestrial biospheres do not appear viable for a variety of reasons – including that they have not worked thus far). Escape, however, represents a loss of humanity and not its preservation. So, even if the possibilities with respect to escape did not belong more in a work of science fiction, they will not be discussed here. A time may well come when those in government initiate programs designed to ensure that some of them survive the effects of catastrophic human-caused environment change. And there will undoubtedly be a politics associated with initiating such projects, funding them and choosing who will be part of government-supported escape.

Ignoring eco-anarchism and green rage may be a more controversial omission, given the extended history of eco-anarchist philosophy. Eco-anarchists can claim descent from Pyotr Kropotkin (1842–1921) and that "the current ecological crisis … makes Kropotkin's work continue to be relevant" (Bowen and Purkis 2018, p. 7). Kropotkin's eco-anarchist ideas were ably taken up and developed by Murray Bookchin and Colin Ward (Bowen and Purkis 2018, p. 7). And, while eco-anarchist ideas were expressed through the Occupy movement (Kinna et al. 2019), their ideas have had limited lasting impact on the conduct of everyday politics. Certainly, some people will feel frustration and anger and may become destructive as a result. But it is wrong to simply equate anarchism with the use of violence (Frazer and Hutchings 2019).

Just as the Occupy movement can be thought to continue an eco-anarchist tradition, Deep Green Resistance, a radical environmental group calling for "decisive ecological warfare" can be thought to continue Manes's Earth First!ers destructive practices in *Green rage: Radical environmentalism and the unmaking of civilization* (Manes 1990). Deep Green Resistance's championing of "both aboveground and underground resistance to industrial capitalism" and acceptance of "sabotage and violence as a necessary tactic" (LeVasseur 2017, p. 109) embraces the "ecotage" that was popularized in Edward Abbey's 1975 book *The Monkey Wrench Gang* (Abbey 1975).

Omitting discussions of eco-anarchism and green rage does not reflect the view that some people will not express anger and frustration at the failure to address catastrophic human-caused environment change and divert from a course toward the final hundred years of humanity. And much of that anger and frustration will be felt and directed toward those in positions of political power and economic privilege. So, the claim here is not that ecotage and green rage will not occur. It is that the likelihood is that the most such actions would achieve is to contribute to catastrophic human-caused environment change and not reduce it.

Violent action might be taken to express the failure of politics, however. This might indicate, however, that (as suggested here) the success of politics lay in the failure to privilege an environmental future over the socio-political status quo, which resulted in catastrophic human-caused environment change and took humanity to the brink of its final hundred years and beyond. Citizens are unlikely to see things this way and will demand that those in government take immediate and significant action. They will call for steps be taken to avert the final hundred years of humanity and demand an act of salvation on the part of those in positions of political power. Radical population reduction is likely to be one of the desperate measures for which people call, though SAI may well become the solution on which all hopes are pinned – a point that will be discussed more fully below.

## Part One: The Politics of Radical Population Reduction

Radical population reduction is an obvious candidate when it comes to addressing the causes of catastrophic human-caused environment change. If it is not possible to stop humans from causing catastrophic environment change, as would appear to be the case from the systematic failure to address the problem, then the next-best strategy to slow and even stop catastrophic environment change would be to reduce the number of humans on the planet.

This 'solution' is not as simple as it would appear to be, however, for important questions arise concerning the communities whose populations will be reduced and how this reduction will be achieved, and both are political questions. An understanding of the politics associated with answering these questions during the final hundred years of humanity population must begin with an investigation of the way that the problem of (over)population is conceived. For this is not a new problem for politics and the history of this 'problem', especially Malthus's famous contribution, provides salutary lessons with respect to the politics associated with the very constitution of the problem of population.

This part is in four sections. Its purpose is simply to raise and reflect on the politics associated with answering the key questions of radical population reduction. It begins with a discussion of the problem of population, using a discussion of Malthus's ideas concerning population control to show how politics underpins the construction of this problem. The second section moves to the politics of the 'who' question: the politics of which people will have their population reduced. The third section offers reflections on the politics of the 'how' question: the politics of how the population is to be radically reduced. The final section is a brief discussion of the modern example of population reduction on something like the scale required to address catastrophic human-caused environment change: China's One-Child Policy.

The politics of radical population reduction has both an inward and an outward face. Its inward face concerns 'our' population (the population of my country). Its outward face concerns 'their' population or populations of 'them' (the populations of other countries). In both cases, one of the crucial questions concerns the control of the designation of the communities targeted for radical population reduction. As will be noted below with respect to the outward 'face' of radical population reduction, foreign aid to facilitate population control or reduction has had very specific targets driven by particular ideological concerns and national interests. A similar effect is likely to occur at a national level, with policies to reduce population reflecting the ideological concerns and the interests of the members of national elites.

The inward face of the politics of population has several significant aspects. In the first instance, the means for achieving radical population reduction are likely to raise various moral problems and result in, what might be thought of as a final round (at a national level, at least) of a contest between religious authority (church) and secular authority (state). Widespread recognition of a crisis will greatly enhance the power of secular authority, and the state is likely to prevail in a contest with the church around the acceptability of measures to control and reduce a country's population through artificial contraception, freely available government supported abortion, voluntary and involuntary euthanasia and a one-child policy. But those in governments in countries of the Global North may struggle to overcome the power of the palliative and aged care industries. These industries will be threatened, in particular, by the introduction of laws to promote voluntary and involuntary euthanasia.

As has been noted, the outward face of the politics of radical population reduction is, for the most part, captured by issues associated with the provision of foreign aid by countries of the Global North to countries of the Global South. One of the features of the politics of the problem of radical population reduction in the context of foreign aid is that it will continue decades of aid allegedly intended to address over-population in countries of the Global South. As a result, the foreign aid provided during the final hundred years of humanity is likely to maintain, where is does not exacerbate, the effects of previous policy aimed at promoting population control in countries of the Global South.

## *The Politics of the 'Problem' of (Over)Population*

The assumption that the problem of catastrophic human-caused environment change is a problem of population, or over-population, is close to automatic. If people are causing the problem of catastrophic environment change, then fewer people must mean less of a problem. While there is undoubtedly a relationship between the number of people on the planet and catastrophic human-caused environment change, it is not a simple relationship and the problem of radical population reduction is a more complicated one than it appears. And an understanding of this complexity requires a more careful consideration of the long history of the constitution of 'the problem of population' in western political discourse, of which the problem of radical population reduction is simply another moment.

Any discussion of the politics of the problem of population, at least in a western context, must begin with Thomas Malthus. The first edition of Malthus's *An Essay on the Principle of Population*, which is the best-known work on this subject, was published in 1798. While he revised it several times over the next 28 years, its central message remains unchanged. This was that any government interference in the natural processes that held populations in check would result in the demise of that society. For Malthus, it was "an evident truth that, whatever may be the rate of increase in the means of subsistence, the increase of population must be limited by it…". This meant that any child whose birth exceeded the number required to keep the population at a sustainable level "must necessarily perish, unless room be made for them by the deaths of grown persons" (Malthus 1826, p. 47).

Malthus railed against religious and moral principles that resulted in government, and other, interventions to prevent starvation. In his view, "we should facilitate, instead of foolishly and vainly endeavoring to impede, the operations of nature in producing this mortality; and if we dread the too frequent visitation of the horrid form of famine, we should sedulously encourage the other forms of destruction, which we compel nature to use" (Malthus 1826, p. 47). Malthus's dire predictions of social disintegration did not fully eventuate. But his ideas have never been forgotten and the final hundred years of humanity is a time at which his grim message will be reconsidered. To uncritically adopt a simple Malthusian position is a mistake in this context, however, as the matter is a more complicated one than such simple approaches would indicate.

The problem with an uncritical and unsophisticated application of Malthusian principles begins with the fact that Malthus expresses the views of a member of a social elite. As a member of that elite he thought immediately of the poor when it came to the lives that were to be sacrificed in the name of maintaining the population at a sustainable level. "Instead of recommending cleanliness to the poor," he wrote, "we should encourage contrary habits. In our towns, we should make the streets narrower, crowd more people into the houses, and court the return of the plague. In the country, we should build our villages near stagnant pools, and particularly encourage settlements in all marshy and unwholesome situations" (Malthus 1826, p. 47).

That Malthus saw the problem of population through the eyes of a member of a social elite may be thought to be less of a problem in the society of his time than it is when it comes to catastrophic human-caused environment change. For such an approach fails to register that the vastly different lifestyles considered normal in different cultures make for very different contributions to the causes of catastrophic human-caused environment change and the extent to which people are predisposed to living an environmentally destructive lifestyle. There can be little doubt that the members of wealthier societies are socialized to adopt unsustainable lifestyles.

According to Ritchie and Roser, per capita greenhouse gas emissions in most of the nations across sub-Saharan Africa, South America and South Asia are "below five tonnes per year (many have less than 1-2 tonnes). This contrasts with the global north where emissions are typically above five tonnes per person (with North America above 15 tonnes)." In many cases, they found, the monthly emissions per capita of those in rich countries exceeded the yearly emissions per capita of those in poorer countries. "The largest emitter, Qatar," they wrote, "has per capita emissions of 50 tonnes per year (1243 times that of Chad, the lowest emitter)" (Ritchie and Roser 2019).

The politics of the problem of population reduction during the final hundred years of humanity begins to emerge when the effects of the elite position of those who define this problem are considered and a discussion of Malthus's thought reveals the extent to which the predispositions of elites condition the conversations around appropriate policy in this context. Ojeda, Sasser and Lunstrum were right, then, to object to an approach (typical of those of the Global North) that "conflates the environmental impacts of human activity with the impacts of human numbers" (2019, p. 2). The problem, for them, "is not that demographic change is irrelevant to environmental harm, but [that] studies have disproven that population growth by itself leads to environmental degradation" (Ojeda et al. 2019, p. 8). The crucial point, when it comes to the 'problem' of population, is that a Malthusian position ignores the fact that "affluence and the related uneven distribution of wealth and political power…, and not sheer human numbers, … [is] the root cause of environmental harm because they lead to overconsumption and overexploitation among a privileged portion of the human population" (Ojeda et al. 2019, p. 8).

So, any discussion of the politics of population reduction during the final hundred years of humanity must involve recognition of the fact that different lifestyles make for very different contributions to catastrophic human-caused environment change. All lives contribute to the problem. Some lives contribute considerably more than others and this merits consideration in the context of the politics of radical population reduction that represents one of the possible, and political interesting, forms of the desperate measures undertaken during of the final hundred years of humanity.

Another of Ojeda, Sasser and Lunstrum's objections to the very form of thinking that underpins discussions of policies associated with radical population reduction is that it involves preventing 'excess' population. The problem with this, as Hartmann, Hendrixson, and Sasser argued, is that it represents "the logic of averted-humans-as-averted-emissions…". This raised "significant questions about the power dynamics of how value in human life is assigned, to/by whom, and at what scale." For, as they pointed out, in this way of thinking "certain lives are deemed investable, others expendable and avertable…" (2016, p. 74). And, as Ojeda, Sasser and Lunstrum wrote, "this kind of calculus reveals the workings of power behind the scenes…" (2019, p. 8).

This is not to deny that the discussion of the policy options in this part reflects this way of thinking and obscure the power working behind the scenes. For there is no doubt that this book suffers from what Ojeda, Sasser and Lunstrum saw as a general problem of hyperbole with respect to the representation of the threat posed by climate change. For this book adopts the, in their view, objectionable position "that life as we know it is threatened at a scale that was, in prior decades, unfathomable; that this is an immediate and overwhelming threat; and that the consequences of not acting expeditiously are untenable." In these "crisis-driven scenarios," Ojeda, Sasser and Lunstrum argued, "strategies that might constrict individual reproductive rights are articulated as a necessary sacrifice" (2019, p. 6).

Further, in Ojeda, Sasser and Lunstrum's view, "a neo-Malthusian lens also opens the door to militarized interventions." For them, such "masculinist approaches to conflict … [also] spin decidedly on (often deeply racialized) gender norms." An effect of this way of thinking, they argued, is that "militarized international strategies for reducing population growth especially in Africa are motivated by fears both of overly reproductive young women and of young men as potential terrorists" (Ojeda et al. 2019, p. 6).

The problem they pointed to is that the position of those who judge that humanity is in crisis (such as in its final hundred years) is delivered from a position of privilege: white, male and middle-class. With the consequence that the way of conceiving the problem and its solution are characterized and constrained by "neo-Malthusian fears, here of overly-reproductive, resource-degrading women and of men who turn to violence in the face of imperiled environments" (Ojeda et al. 2019, p. 6).

The preceding quotation from Ojeda, Sasser and Lunstrum is an important one in that it makes clear that, no matter how objective the problem of population may seem to be, the way it is understood is crucial for how it will be addressed. Much of the fear of population amongst those in countries of the Global North is of a rampant population explosion in the countries of the Global South. The power of those in countries of the Global North will mean that this mentality is likely to apply in practice. This is despite the fact that this is a misrepresentation of the situation because it reduces the problem of population to a question of numbers of people, rather than to that of numbers of people and the lifestyles they lead. In this way, the politics begins to insinuate itself into the process of addressing the problem of radical population reduction.

## *The Politics of the 'Who' of Radical Population Reduction*

This discussion of the politics of the problem of (over)population has already begun to lead to the question of who will be seen to be 'excess' (those who are to be 'averted'). As has been discussed, when Malthus viewed the problem of (over)population it was obvious to him that the poor were 'excess'. That his view reflected his position as a member of an elite meant that he was oblivious to the politics of his answer to the 'who' question. Those who dominate the 'conversation' concerning the problem of population will determine its solution and a crucial aspect of that solution concerns which population in a society or on the planet will be targeted for radical reduction.

Both the inward and outward faces of the politics of radical population reduction become important when answering the 'who' question of radical population reduction during the final hundred years of humanity. A complicating factor that affects both faces of the politics of radical population reduction is that population size is associated with national prestige in the eyes of some political leaders and their followers. Indeed, growth across a variety of measures, especially the size of the economy and gross domestic product (both of which are related to population), is often taken to reflect a nation's success and status. Those seeking to introduce policies to achieve radical population reduction will face a challenge, as some leaders and their people take a decreasing population to be a marker of decline and a decreasing population is likely to result in reductions in the size of the economy and falls in GDP.

Korkut and Eslen-Ziya provide an insight into this aspect of the politics of the 'who' of radical population reduction in their 2016 article on population discourse in Turkey. After noting that "sustaining a growing population is seen as politically crucial across many political contexts", they "confront the puzzle that a discourse supporting the increase of population has emerged in a context where population growth and fertility rates are robust…" (Korkut and Eslen-Ziya 2016, p. 557). Their explanation for this phenomenon is that the "alarmist discourse about population," reflected "fears of other kinds of social change on the part of the governing Justice and Development Party (AKP), representing political Islam, and President Recep Tayyip Erdoğan…" (Korkut and Eslen-Ziya 2016, p. 557).

Like population politics in Russia and Hungary, Korkut and Eslen-Ziya argued, "Turkish population politics makes more sense if we consider the importance of childbearing in the national identity and the economic value of population growth for political regimes" (2016, p. 558). As Gökarıksel, Neubert and Smith put it, "Erdoğan locates Turkey's future strength in a young and large population, stoking fears of population decline…" (2019, p. 561). But there are other drivers of Turkey's leader promotion of population growth, Korkut and Eslen-Ziya added, Turkish "politicians' emphasis on having more children is a reflection of conservative identity politics where 'good mothers' serve the social 'order,' a tool of patrimonial power" (2016, p. 558).

Along similar lines, in discussing populists in India (Vijaykant Chauhan), the US (Charlottesville protestors) and Turkey (Erdogan), Gökarıksel, Neubert and Smith identify logics that "motivate and support right-wing movements across the globe that see their hegemonic positions threatened by migration and the reproductive fertility of racialized others" (2019, pp. 555–6). As Gökarıksel, Neubert and Smith put it, populist movements express a "fever dream" that "implies an orientation toward the future… that is demographically apocalyptic for the dominating population and thereby requires active, often violent intervention" (Gökarıksel et al. 2019, pp. 555–6).

For the members of a dominant group "biological survival becomes the stake in an intimate geopolitical battle, where declining demographic dominance is read as 'a symptom of broader decay' and future success depends on demographic growth and meticulous engineering of the population to ensure dominance" (Gökarıksel et al. 2019, pp. 556). While migration is one cause of a decline in the dominant group's demographic dominance, reproduction by members of non-dominant groups is also a threat. These threats, according to Gökarıksel, Neubert and Smith "are distinct but also linked and sometimes blurred: in fact, often the intent is to blur, such that people who have lived in a place for generations are eternal migrants, and their reproduction a form of invasion" (2019, pp. 556).

This way of thinking relies on very powerful populist fictions. The most important of these is the projection of a 'common folk' or 'ordinary people'. This is a political construct that has no real-world referent. As Elchardus and Spruyt pointed out, "the 'ordinary people' appear less as a specific group that can be delineated on the basis of socioeconomic or sociocultural characteristics, [and] more as a concept that primarily refers to a certain way of thinking, to a set of perceptions and propositions about society and politics" (2016, p. 114). Crucially, the construct of 'ordinary people' creates a division between those who can legitimately participate in processes of determining who 'ordinary people' are and what they want and those who cannot participate. For Elchardus and Spruyt, this "way of thinking – common sense – is established not only as the desirable way of thinking, but also as the only truly democratically legitimate foundation for policy. In that way populism is able to structure social and political space, divide it between us and them, between friend and enemy" (2016, p. 114).

As was discussed above, populism, in most of its forms, requires policies that promote growth, including population growth. Here we encounter the other side of the populist political coin: that policies to promote radical population reduction could be rejected because they accept and even promote decline. In this case, the 'ordinary people' reject a political elite whose members they do not resist, and even encourage, decline (with the suspicion that these policies of decline serve the interests of members of the elite). Policies designed to radically reduce a specific national sub-population simply "emphasize the distance between we the ordinary people and the elite…" (Elchardus and Spruyt 2016, p. 117).

Policies to promote radical population reduction will meet resistance because 'ordinary people' see them as overt acceptance of decline and a loss of the way things were. Populist believe that enough decline, especially in their way of life. For, "nostalgia for a lost heartland and the perception of decline has occurred in the present day are close to the core of populism. … a negative view of the state of society – declinism – … [is] strongly and positively related to populism" (Elchardus and Spruyt 2016, p. 117). For the populist, the only acceptable decline is that is the number of 'them' – which may refer to a 'them' that has infiltrated their nation or a 'them' in other countries (in particular, those of the Global South).

Studies of previous aid programmes designed to produce population control in countries of the Global South also point to a variety of political aspects to radical population reduction. Of course, advocates of contemporary population reduction programs developed and delivered with respect to addressing catastrophic human-caused environment change claim that they will avoid the problems of population control's 'dark past'. Critics, however, are clear that even though these advocates cast contemporary population control in terms of social justice and environmental protection they continue to return its 'dark past'. For these critics, while "advocates attempt to elide the racialization inherent in their current narratives about burgeoning populations in the global South as environmental degraders… [their] narratives continue to inscribe bodies in the global South as those in need of intervention, typically with family planning, presented as an environmental good" (Hendrixson 2019, p. 788).

Recognising the continuity of policies designed for population control as means for achieving development and radical population reduction and those developed with respect to catastrophic human-caused environment change and during the final hundred years of humanity is important. For Hendrixson, attempts to deny continuity between past policies for population control for development and present and future policies of radical population reduction for environmental protection were "similar to confusions in post-colonial thinking and politics, where the prefix 'post' presupposes that colonial domination ended and a new era of independence began." In this view, the thinking that underpinned population control for development continues in population reduction for environmental protection and "manifests in global programmes, country population policies, and neo-Malthusian ideologies in the context of FP2020 [Family Planning 2020] and the SDGs [Sustainable Development Goals]" (Hendrixson 2019, p. 800).

Another criticism of past population control programs relates to to high levels of coercion associated with many of those programs. In reviewing studies of sterilization programs run by foreign aid organizations in Bangladesh, for example, Hendrixson found "that it was mainly the poor who were sterilized; payments did act as incentives for high proportions of the poor; paid 'motivators' targeted the poor who had a financial reason to undergo sterilization; and many people regretted accepting sterilization" (Hendrixson 2019, p. 792).

The principal problem identified by critics of population reduction programs in countries of the Global South that were run as development assistance from countries of the Global North is that they served the interests of the donor country or countries. Bendix, for example, offered a revealing account of the switching of German population programs in Tanzania from population increase for development to population decrease for economic stability. Thus, whereas "German colonial intervention to increase population was explicitly legitimized with reference to economic interests; contemporary German policies in the field of population and reproductive health are primarily presented as philanthropic, altruistic endeavours." But the overall goal, for Bendix, remained the same: "to transform Tanzania into a 'stable', 'functioning' society with which Germany could engage in profitable economic exchange..." (Bendix 2016, p. 65).

In both cases, "interventions into reproductive health and reducing population growth in Tanzania" were deemed necessary (Bendix 2016, p. 65). This aid follows a general pattern in which "population policies in the Global South serve countries of the Global North through the 'containment of a superfluous or redundant labour force, the maintenance of political stability, and the perpetuation of dependent social relations'" (Bendix 2016, p. 65). Bendix concluded that this meant that "development assistance can no longer be theorized as a strategy to fight 'surplus population' in the Global South ..., but rather as more generally serving the regulatory interests of the North or global capital" (Bendix 2016, pp. 65–6).

While radical population reduction brings some to think of the way population control was used as part of colonization, for those who know its history, discussions of radical population reduction also recall the 'science' and politics of eugenics. A variety of eugenics movements arose across the world soon after the turn of the twentieth century. In the US, the American Eugenics Society (AES) "and its religious allies supported immigration restriction and involuntary sterilization and generally promoted positive eugenics..." (Wilde and Hopkins 2018, p. 18). The failure of positive eugenics, encouraging 'desirable' (i.e., white Christian) people to have more children, meant that "the AES and the religious leaders affiliated with it turned to 'negative eugenics' and began pushing for the legalization of birth control" (Wilde and Hopkins 2018, p. 18).

Various factors, but principally the Nazi's promotion of eugenics, meant that its supporters were less likely to refer to it by name, according to Wilde and Hopkins, however, "much eugenic thought, and activism around birth control, remained but with two differences" (2018, p. 18). The first difference was that, "although the focus was still on poor people of color, instead of being concerned about non-white immigrants' fertility in the USA, activists became focused on fertility in the developing world, and, to a lesser extent, African Americans in the inner cities" (Wilde and Hopkins 2018, p. 18). The second difference was that "eugenicists began to engage in a more 'discreet and mild-mannered form of eugenics' where they attempted to accomplish 'eugenic control' through 'population control'" (Wilde and Hopkins 2018, p. 18). Another interesting feature of the quieter form of eugenics, of which Wilde and Hopkins wrote, was that "eugenicists strategically promoted population control as crucial to preventing 'the imminent destruction of human society'…" (2018, p. 18).

Several significant concerns arise with respect to their policies intended to achieve radical population reduction in relation to answering the 'who' question when it comes to the final hundred years of humanity. The evidence is clear that people of countries of the Global North make the greatest contribution to the causes of catastrophic human-cause environment change. The evidence is also clear that it is unlikely that the radical reduction of populations undertaken during the final hundred years of humanity will be that of the radical reduction of the populations of the countries of the Global North. These factors mean that we can expect continuity with respect to how the radical reduction of the global population is achieved: from population control for development to radical population reduction to address the effects of catastrophic human-caused environment change. The outward looking face of radical population reduction, from the postition of someone from the Global North, involves 'us' assisting 'them' to reduce 'their' population – probably as an act of great generosity. The outward face also involves maintaining a national population in order to retain prestige in the eyes of those of other countries.

The inward version of this face is the defence of 'our' population. This might be a national population but is more likely to be a demographically identifiable population. A decline in this population becomes a marker of the loss of status and prestige of those who once belonged to a dominant group. In the populist version, the population that is threatened by decline is 'ordinary people' who must defend their numbers from attempts to reduce it by members of the political elite. This populist inflection may become hard to differentiate from eugenics. In the end, though, the answer to the 'who' question of radical population reduction during the final hundred years of humanity will be determined by those with the economic, political and discursive resources that enable them to direct policy and provide that answer in practice.

## *The Politics of the 'How' of Radical Population Reduction*

One of the consequences of Malthus determining that the poor represented the excess members of his society was that also resolved the problem of the 'how' of radical population reduction. Being impoverished meant that the poor were much more likely to fall victim to starvation, disease and violent death at the hands of others. The question of who was excess answered the question of how their population was to be reduced. Making the poor's living conditions worse would help to promote disease, but no definitive action is required from those in government (indeed, encouraging inaction in the face of demands to assist the poor was Malthus's principal goal).

That the poor are also the most vulnerable to the effects of catastrophic human-caused environment change means that they represent the easiest targets when it comes to the 'how' of radical population reduction and denying them resources is the simplest answer to the 'how' question. But even those who govern the most ruthless countries of the Global North might find active population reduction to be too controversial, at least in the early period of the final hundred years of humanity. So, they may well allow 'natural' processes to take care of the excess populations in poor countries by denying aid to the people of those countries. While those in governments who can accept decline will confront the 'how' question when they look inward to doing their share by reducing their national population as quickly as possible.

Two questions arise when first considering the politics of the 'how' of radical population reduction. The first is 'will it work?' The second is 'how will it be done?' The first question is important only in that it bears upon what is truly required for radical population, which is likely to different from what is done – at least in the early period of the final hundred years of humanity. The second question is central, in that the means for radical population present moral or ethical challenges that will be confronting for most other members of society, but will force resistance from leaders of Christian and Muslim sects.

While defending such an outcome is difficult, it is important to bear in mind that radical population reduction is not simple to achieve. Bradshaw and Brook stated the obvious when they wrote that "amoral wars and global pandemics aside, the only humane way to reduce the size of the human population is to encourage lower per capita fertility" (Bradshaw and Brook 2014, p. 16610). The problem with this more comfortable approach is that it is unlikely to have a great enough effect in time to help reduce catastrophic human-caused environment change. Bradshaw and Brook estimated that, even if every country in the world had adopted a one-child policy in 2014, the best that could be hoped for was to maintain the world's population at 2014 levels by 2100 (2014, p. 16610).

In their view, "even a catastrophic mass mortality event of 2 billion deaths over a hypothetical 5-y window in the mid-21st century would still yield around 8.5 billion people by 2100" (Bradshaw and Brook 2014, p. 16610). The most important points Bradshaw and Brook made in their article is that there are "no easy policy levers to change the size of the human population substantially over coming decades, short of extreme and rapid reductions in female fertility…" (2014, p. 16610). That is, radical population reduction would have to be very radical to be effective. The politics of its inward face reflects the level of the moral or ethical challenges that policies to produce radical population reduction, even ineffective ones, represent. The politics of its outward face return us to the asymmetries of the relationship between countries of the Global North and those of the Global South.

While voluntary contraception, abortion and, where possible, euthanasia may contribute to radical population reduction, mandatory forms of each will be necessary if it is to be achieved. This raises the specter of forced sterilization, abortion and euthanasia and presents a dystopian image of an inexorable bureaucratic machine refusing to respect human life and, in countries where they were once recognized, individual rights, as it methodically works to reduce the world's population. While, as Bradshaw and Brook have pointed out, policies to reduce the world's population will not have a profound effect, the introduction of these policies will meet resistance on moral or ethical grounds, particularly from some Christians and Muslims.

Mandatory abortions, involuntary euthanasia and forced sterilization represent the most draconian of the measures that might be introduced. These are controversial policies that raise a myriad of moral or ethical issues. While these policies will challenge everyone, those with religious authority are most likely to oppose them claiming they contravene moral, ethical and spiritual principles that are central to their faith. The principal form of the politics of the 'how' of radical population reduction, then, is the return of a conflict between political and religious leaders.

Competition, perhaps, conflict, between church and state has long been a feature of the politics of western societies. An individual's right to follow her or his own religious principles in defiance of secular authority was important to the development of liberal societies (whose members often, somewhat inaccurately, trace the principle to the Peace of Westphalia). As part of the emergence of a liberal politics, religion was construed as private and purely a matter of individual conscience, which "led to the complete elimination of any reference to values and explicit religious content from those areas of social co-existence regulated by state laws" (Ferrari 1988, p. 535). The separation of church and state protects the individual from political interference in their spiritual practices. For many Christians living in liberal-democracies, then, policies to produce radical population reduction will also represent an assault on people's individual rights.

The problem will be different for Muslims, especially those in non-Western countries, whose culture is neither as highly individualistic nor oriented toward a profound division between church and state. For significant numbers of Muslims, a clear separation between religion and politics is not meaningful. According to a poll conducted by the Pew Research Centre in 2013, many Muslims wanted "religion to play a prominent role in politics" and an "overwhelming percentages of Muslims in many countries want Islamic law (sharia) to be the official law of the land" (Pew 2013). Care must be taken here, for the experience "in Iran, which arguably boasts the world's only Islamist government, [is that] clerical governance has led to a steep decline in religious observance; in 2011, the Iranian Ministry of Culture and Islamic Guidance lamented that after more than 30 years of theocratic rule, only three percent of Iranians attended Friday prayers. (Prior to the revolution, the figure was almost 50 percent.)" (Ruthven 2016). There may be circumstances, then, when the lack of separation of church and state can become a problem for some Muslims.

An initial question that will affect both Christians' and Muslims' response to morally or spiritually objectionable policies designed to reduce the population will concern whether catastrophic human-caused environment change is taken to be a manifestation of divine will or of human pride. The latter could refer to the failure to care for the planet God gave them (and for which God may have given them responsibility to protect and preserve); or it could refer to people's lack of humility in thinking that they and not God will determine how long life on Earth persists. The situation is further complicated by the fact that, as was discussed in Chap. 3, some Christians and Muslims believe that the end of humanity is foreordained and manifests divine will.

Neither Christianity nor Islam expresses a single position with respect to radical population reduction. While some sects and religious leaders in both faiths reject it, others do not. The only thing that might be said of all religions is that their leaders want more followers rather than fewer followers and that this would generally imply that they would prefer that people of their faith procreate. Support for a general resistance to policies intended to reduce a population might be found in a biblical injunction (Genesis 1:28) to: "Be fruitful, and multiply…" (King James Bible Online). While, in Islam, according to Bachelor, "a religious duty to produce a large *ummah* [community]" follows from Muhammad call's to "Marry and multiply for I will be proud of you on the Day of Judgement before the other nations" (Batchelor 2019, p. 27).

Christian and Islamic conceptions of the soul differ significantly, with Islam having a developmental view of the soul, and this will affect their responses to some of the policies require for radical population reduction. Whereas, for Christians, the soul is the soul, for Muslims, the soul progresses through three stages: from a selfish soul (*nafs ammara bissu*), through a self-blaming soul (*nafs lawwama*) to a higher form (*nafs mutmainna*) (Ahmed 2011). When it comes to radical population reduction, the Christian must consider her or his "eternal essence (the soul) and a relationship to God that transcend physical reality" (Bratton 1992, p. 102). It is more complicated for Muslims. Nonetheless, notions of immortal souls and divine essences complicates the situation when it comes to radical population reduction.

For followers of both faiths, however, the crucial question with respect to the politics of radical population reduction during the final hundred years of humanity concerns the extent to which the earthly is superseded by the divine. For the Muslim seeking the development of the soul and the Christian seeing its full expression, a "willingness to adapt Christian [and Muslim?] values in population ethics will depend at least partially on whether one believes there is a resurrection and whether one believes our final state depends on our relationship to God rather than on the number of our progeny" (Bratton 1992, p. 102).

Both Christians and Muslims accept natural contraceptive methods. Some Christian sects accept artificial contraception. A similar situation holds for Islam. For some Islamic scholars, the Quran's silence with respect to contraceptive use is intentional. According to Ataullahjan, Mumtaz and Vallianatos, different sects' positions on contraception is method dependent. "Most Islamic scholars agree that irreversible methods such as sterilization are impermissible, while reversible methods (i.e. condoms, injectables, pills, IUDs, etc.) are permissible" (Ataullahjan et al. 2019, p. 50).

But even when they agree about the acceptability of a method for contraception, they may disagree as to whether it ought to be used. For some, "limiting fertility is a challenge to predestination (*qadr*) and reliance on God (*tawakkul*)... with many viewing contraceptive use as an affront to God as the ultimate provider" (Ataullahjan et al. 2019, p. 50). Ataullahjan, Mumtaz and Vallianatos suggested that Islam was tacitly pronatalist, adding that "some scholars have compared contraceptive use to infanticide. Others cite demographic concerns and conspiracies to support their resistance to contraceptive use" (2019, p. 50).

The pressure to retain followers remains important, however. Public opposition to policies that have been deemed necessary due to catastrophic human-caused environment change could cause religious organizations to appear out-of-date or somehow wrong when increasing numbers of people believe that it is imperative that those in government address those effects. A public stand against these policies could, applying the concept of 'salient controversy', result in people turning their backs on religious organizations. The danger of losing adherents and alienating potential followers might cause religious leaders to withhold criticism of, if not provide support for, policies that they might previously have opposed.

Djupe, Neiheisel, and Conger studied the relationship between a rise in those who indicated they had no religion during the US Census and the passage of anti-same sex marriage legislation by some State legislatures. In doing so, they found, legislators "created a set of conditions ripe to make the Christian Right salient to the public in some states more than others" (Djupe et al. 2018, p. 918). This salience, however, led to a loss of support for Christianity in the communities in which anti-same sex marriage laws were passed. These moves, which began in 2004, resulted in an increase of between 2% and 8% of people indicating that they had no religion in the Census. According to Djupe, Neiheisel, and Conger, 29 states had introduced a ban by 2010. "These states were more likely to be evangelical and had smaller proportions of nones in them in 2006, but by 2010 that gap between the nones in marriage ban states and those in states with no marriage ban had been cut in half (it went from 3.1 percent to 1.4 percent over that period)" (Djupe et al. 2018, p. 918).

The crucial question that arises in this context relates to the numbers of those who turn their backs on the church and of those who embrace it because of catastrophic environment change bringing humanity to the brink of and beyond its final hundred years. One aspect of the politics around policies with respect to radical population reduction will bring the leaders of some Christian and Islamic 'churches' into conflict with political leaders seeking to implement policies directed toward radical population reduction, many of which appear to infringe the spiritual and moral principles articulated in some Christian religions, or by the leaders of those religions. As Djupe, Neiheisel, and Conger's research suggests, the churches run the risk of losing adherents if they adopt policy positions that overtly contradict a dominant popular sentiment.

## *China's One-Child Policy*

Before leaving a discussion of the 'how' of radical population reduction, a brief discussion of China's one-child policy (OCP) provides some assistance in working through the issues that arise with respect to radical population reduction. For the OPC, which the Chinese government adopted between 1979 and 2016, provides an illuminating example of a large-scale attempt at population control, and radical reduction of the world's population would have to reproduce many aspects of the OCP. The questions that arise for those considering radical population reduction concern the success of the policy and the unintended effects of the policy.

While a desire to promote economic development led the Chinese government to abandon its OCP in 2016, the Chinese government has long claimed that policy was successful in controlling the country's population. This policy began to take effect in 1980, with Chinese government officials claiming that the policy prevented 250 to 300 million births and, according to the census conducted that year, kept China's population at 1.27 billion in 2000 (the government's goal had been to keep the population to 1.2 billion by 2000) (Hesketh et al. 2005, p. 1172). If trust can be placed in the statistics produced by the Chinese government, which like all governments does not want to be embarrassed by evidence of policy failure, then the policy was a success (only later being changed to promote consumption and boost the economy).

That the policy was the success the Chinese government claims that it was is contradicted by several factors. The first is that the decade prior to 1980 saw a more significant decline in the fertility rate (from 5.9 to 2.9). The rate continued to decline after the introduction of the OCP, though, and stabilized at around 1.7 after 1995 (Hesketh et al. 2005, p. 1172). Another confounding factor is that as fertility rates fell in China they also decreased, and to even lower levels, amongst China's East Asian neighbors; falling to 1.04 in Singapore, 1.38 in Japan, and 0.91 in the Hong Kong Special Administrative Region. As Hesketh et al. suggested, "the policy itself is probably only partially responsible for the reduction in the total fertility rate" (Hesketh et al. 2005, p. 1172).

One problem with the OCP was that its effects on the make-up of the reduced population could not be fully predicted and controlled. As Wang and Zhang argued, "we cannot conclude that such population control policies will achieve the goals of policymakers, who cannot control which segment of the population is reduced" (2018, p. 233). The selection problem that Wang and Zhang found to have resulted from China's OCP was an increase in the population of rural areas relative to urban areas, with even more boys, relative to girls, being born in rural areas than in urban areas. This resulted from the fact that parents in rural areas were less vigorously policed with respect to the OCP, which meant many more had two children, and from the fact that they had an even stronger preference for sons than their urban counterparts. Thus, "given that son preference is more prevalent in rural areas than in urban areas, the sex ratio is much more distorted in rural areas (having more boys than girls) due to sex selection under the OCP" (Wang and Zhang 2018, p. 232).

Other authors, who Wang and Zhang cited, claimed that the OCP produced significant personality differences between Chinese sons in rural and regional areas. "Cameron et al. even argued that the OCP produced significantly less trusting, less trustworthy, more risk-averse, less competitive, more pessimistic and less conscientious 'little emperors' in urban China" (Wang and Zhang 2018, p. 232). Though Wang and Zhang were themselves provocative in suggesting that "the OCP reduced the average human capital level of the next generation…" (Wang and Zhang 2018, p. 232).

Of course, the most commented upon aspect of the OCP was the level of coercion that was required for its implementation. As Goodkind put it, "both friends and foes of that intervention might reconsider the 'awesome Strength' of China's 'longest campaign', which was 'coercive in central design and intent' and required parents to make 'coerced choices'" (2017, p. 1394). Novelist Ma Jiang argued that the OCP meant or allowed the Chinese government "to intrude into the most intimate aspects of an individual's life, stunting relationships, destroying traditional family life and spreading fear" (Ma 2013).

The result of the OCP, in Ma's view, was that women "lost sovereignty of their bodies. The state owns their ovaries, fallopian tubes and wombs," he wrote, "and has become the silent, malevolent third participant in every act of love" (Ma 2013). He also wrote of an incident in 2007 in which pressure from those in higher levels of government to meet birth targets caused local officials in Bobai County to launch "a vicious crackdown on family-planning violators. Squads had rounded up 17,000 women and subjected them to sterilisations and abortions and had extracted 7.8m yuan (£800,000) in fines for 'illegal births', ransacking the homes of families who refused to pay" (Ma 2013).

Greenhalgh, on the other hand, saw the consistent representation of the OCP as a manifestation of a violent dictatorial regime as an effect of a distinct, geopolitically important discourse. In her view, it was one element of "a particular Cold War narrative of China as the 'coercive (backward, unscientific) other' to the 'free (modern, scientific) America' … (Greenhalgh 2018, p. 722). Greenhalgh did not deny "that coercion was an important part of policy enforcement in certain times and places…" (2018, p. 722). However, she argued that to make "coercion the single most important thing to know about China's population politics…" was overly reductive and even anachronistic. "The coercion critique is important, even today," she wrote, "but it needs to be supplemented by broader understandings of the ways population governance itself has been transformed with China's deepening modernization and globalization" (Greenhalgh 2018, p. 722).

Greenhalgh's work, then, suggests that there is a politics to the representation of China's OCP that may play a role in promoting resistance to policies to achieve radical population reduction during the final hundred years of humanity. But even Greenhalgh acknowledged that the implementation of the OCP involved significant levels of coercion (and, for that matter, corruption – which may also arise when it comes to policies to produce radical population reduction). The question of whether the OCP was a success may be thought the greater issue in this context. But, even this does not fully resolve the question of the desirability of measure to achieve radical population reduction, as one of the downward pressures on population seems to have been increasing affluence (see Hesketh et al. 2005, p. 1172).

Policies that have been designed to radically reduce the world's population are likely to have to mirror, and may even need to be more extreme than, those adopted under China's OCP. So, they will raise a variety of significant political problems. Most importantly the problem of (over)population has a Malthusian source that introduces troubling overtones to the discussion. But even without their Malthusian history, policies to achieve radical population reduction raise important political questions concerning its 'who' and its 'how'. The 'who' question leads to a focus on the relationship between population and national prestige, the populist inflection that these policies could readily take on (assuming that they are not an inherent part of such policies) and previous experience with population control programs that were intended to promote development in countries of the Global South. While, in addition, salutary lessons are offered by a consideration of China's One-Child Policy.

## Part Two: The Politics of Stratospheric Aerosol Injection

The first section in this chapter addresses the question of whether this part should have been written. Those who object to discussing possibilities with respect to all forms of geoengineering, of which stratospheric aerosol injection (SAI) is one, argue that, in the first instance, such a discussion shifts attention from the problem of reducing greenhouse gas emissions and may even encourage people, firms and governments to treat geoengineering as the solution to the problem of 'climate change'. Other problems with discussing possibilities with respect to any form of geoengineering are that it reproduces a 'technology fix' mentality that represents a refusal of responsibility and a failure to address human beings' relationship with nature. A third problem with discussing any form of geoengineering is that such a discussion begins with the projection of catastrophe or crisis. While this is also taken up in the first section of this part, this book is a projection of catastrophe or crisis and succumbs to this problem.

The reasons not to discuss SAI are valid. But this book has laid out an argument as to why nothing has been done and why nothing will be done until circumstances begin to threaten the status quo and something needs to be done to protect the social order governments have worked to preserve. Indeed, as Gunderson, Stuart, and Petersen suggested, "many more may be attracted to geoengineering because it envisions a pathway that maintains the current social order" (Gunderson et al. 2018, p. 15). By this point, though, humanity will already be in its final hundred years.

This will give geoengineering in general and SAI in particular considerable policy salience. So, given that SAI will be considered an option for addressing catastrophic human-caused environment change, the second section explains SAI. It begins with a discussion of geoengineering, which includes both solar radiation management (SRM) – of which, SAI is one form – and carbon dioxide removal (CDR). While CDR is a better means for addressing the effects of increasing greenhouse gas emissions, it will be much slower to affect the average global surface temperature and it will be more expensive. Some CDR will be undertaken, but SAI is likely to be the principal means adopted to address the effects of catastrophic human-caused environment change. So, this part finishes with an explanation of SAI.

SAI raises at least seven questions that have a political aspect. The first concerns research and development. Who conducts the research, how it is conducted and who evaluates it are all important political questions that arise with respect to SAI. Other questions concern the provision of SAI, its monitoring and assessment, its regulation and its maintenance (abrupt cessation of SAI is likely to have disastrous effects). Given that SAI is likely to have a deleterious effect on some countries, questions of compensation arise with respect to SAI. Even if compensation is available and especially when it is not, some of those in countries harmed using SAI might resort to reversing its effects through counter-geoengineering. Such a scenario suggests the possibility of weaponizing SAI.

## The Politics of Discussing Geoengineering

Three problems have been identified with respect to discussing any options with respect to geoengineering (other, perhaps, than painting surfaces white). The first is that to discuss any form of geoengineering will cause a shift of attention from reducing greenhouse gas emissions to sustainable levels to a 'solution' that does not require any change of behavior. The second problem, which is related to the first, is that taking up most forms of geoengineering, including SAI, does not change the values and mentalities that resulted in catastrophic human caused environment change and have led to the maintenance of a trajectory toward the final hundred years of humanity. The same instrumental exploitation-oriented view of nature will persist, and at best, the end of humanity may be delayed but never prevented because human beings' relation to nature remains unchanged.

The third problem with discussions of geoengineering is that they are discussions of a 'last resort' and this implies that the situation is catastrophic. The problem with catastrophes is that they can produce, what Carl Schmitt referred to as, a 'state of exception' in which normal rules of governance and perhaps even decency are suspended in order to deal with the catastrophe. Given that the premise of this book is that such a state of exception will eventuate, and that the loss of decency is the loss of humanity, this problem may be taken to be is the problem of this book.

Suppression of discussion of geoengineering was, and for some is still, justified on the grounds of it being a 'moral hazard'. Here the phrase refers to the possibility that knowledge of the existence of other ways to address the effects of the increasing concentration of greenhouse gases than that of reducing greenhouse emission will result in little effort being made to reduce emissions. As Merk pointed out, though, the problem is not a 'moral hazard', as the phrase is normally used. "Here the problem is more akin to a 'lack of self-control' or an escape from moral responsibility' (Merk 2019, p. 136). As has been noted, the problem identified here is that the possibility that knowing about geoengineering frees people and those in firms and governments from having to reduce their carbon footprint. This is particularly true of "those already opposed to emissions reductions [who] will use SG [solar geoengineering] as yet another excuse not to act" (Merk 2019, p. 137). And a study by Campbell-Arvai and colleagues found that knowing about carbon dioxide removal (CDR – one form of geoengineering) can reduce the perceived risk of climate change… [and] was associated with reduced support for climate mitigation policies, and this effect was most pronounced for those with a conservative political orientation" (Campbell-Arvai et al. 2017, p. 334).

Putting geoengineering on the political agenda is no less contested than raising it as a scientific possibility, which has been a crucial factor in geoengineering not having reached the political agenda. For politicians and policy makers the problem is that they will face "accusations of bad faith if they promote Plan B when it is they who are in a position to implement the superior Plan A, cutting greenhouse gas emissions" (The Bulletin of the Atomic Scientists 2014, p. 18). The result of this reluctance on the parts of politicians and policymakers means that, according to Lin, even in 2019 the bodies that addressed the problem of catastrophic human-caused climate change were yet "to take a position on – or explicitly acknowledge – the controversial subject of solar geoengineering" (Lin 2019, p. 125). Even discussing geoengineering has a politics.

Another problem with discussing geoengineering is that is manifests a 'techno-fix' mentality in which science and engineering are relied upon to solve humanity's problems, when others believe that this faith in science and engineering is the reason humanity is facing its final hundred years. As Gunderson, Stuart, and Petersen put it, geoengineering represents "the ultimate techno-fix. Portrayed as a means to control nature, geoengineering may seem like the ultimate techno-scientific achievement and the rational way forward" (2018, p. 12–3).

Thiele described the mentality of geoengineering as Promethean. It "posits humanity as an exceptional, inventive species possessing scientific knowledge and technological capacities that allow the beneficial modification of its environment" (Thiele 2019, p. 461). Prometheans even express "*excitement* about the prospect of using solar geoengineering to transform the Earth" and "delight in the prospect of using technology to create a new Earth…" (The Bulletin of the Atomic Scientists 2014, p. 25). Opposed to this mentality, in Thiele's analysis, is a Gaian humility that "seeks to harmonize us, as one among many interdependent species, with the natural world." For Gaians geoengineering "is a dangerous act of hubris aimed at gaining full-fledged mastery over the planet" (Thiele 2019, p. 461).

Rather than see geoengineering in terms of a continuation of a problematic relationship between humans and nature, however, Honegger construed it as establishing "a new relationship between human beings and their environment…" (Honegger 2019, p. 79). But this new relationship may be an ever more reductive and instrumental view of nature in which human domination over nature is celebrated. So, from this perspective, discussions of geoengineering require new voices, as it "poses spiritual and cultural challenges that may require the voices of spiritual thought-leaders and non-experts to be added to previously enumerated perspectives" (Honegger 2019, p. 79).

The most troubling problem with any discussion of geoengineering, however, is the fact that it relies on a catastrophe or crisis narrative. It is generally presented, as it is being presented here, as a last resort. The only difference between this and other analyses is that catastrophe is treated as a given and a state of exception is assumed to be inevitable due to the factors that prevent action to address catastrophic human-caused environment change. Carl Schmitt laid the groundwork for the concept of 'state of exception' that is applied here. In brief, the 'state of exception' refers to a situation in which those in positions of (sovereign) political power could determine that normal rules of public life did not apply.

This is not an 'objective' determination and is at the discretion of those in positions of political authority. It is a moment when those in government decide to go 'outside' the law. "The exception, which is not codified in the existing legal order, can at best be characterized as a case of extreme peril, a danger to the existence of the state, or the like. But it cannot be circumscribed factually or made to conform to a preformed law" (Schmitt as quoted in Gray 2007, pp. 179–80). Once those in positions of political power decided that a state of exception exists, then they were free to do what they deem to be required to return the situation to normal (Gibbs 2019, pp. 73–4).

The lesser, though significant, fear that arises in these circumstances is that declaring a 'state of exception' results in climate technologies being "taken out of the realm of normal political discourse and debate and into the realm of 'necessity' and urgency, and the rest of climate change therefore becomes subject to security logics…" (Corry 2017, p. 310). A greater source of apprehension is that the declaration of a 'state of emergency' initiates "a process of political degeneration in which rights are suspended, dissent is suppressed, power is concentrated, and decision-making is rushed and myopic" (Horton 2019, p. 37). The greatest fear is that the public will demand unilateral action by those in their government. For the world of catastrophic human-cause environment change is one "in which publics will demand action, and leaders will want to show they are responsive. A leader with fragile public support and few restraints on power may look to the short-term gains from unilateral deployment [of geoengineering]: being seen as taking action, and possibly even succeeding in ameliorating climate-change impacts" (Victor 2019, p. 43).

## What Is SAI?

Despite these legitimate concerns about raising it as a possibility, geoengineering, in the form of SAI, will be the subject of the discussion in the remainder of this part. For, as Victor put it with respect to all forms of geoengineering, "formerly a freak show in otherwise serious discussions of climate science and policy, geoengineering today is a bedfellow, albeit one of which we are wary. The option may be needed as a hedge against unexpectedly harsh changes in climate" (Victor 2008, p. 323). But, given the tenor of the discussion in the first part of this book and the politics associated with introducing, regulating and maintaining it, SAI merits careful consideration. Pasztor framed this in term of 'ifs', rather than 'whens', so the urgency is greater than he suggested. But he is right to argue that, "if current predictions about our inability to address rising temperatures quickly enough turn out to be accurate, the global debate over the availability and feasibility of climate geoengineering is about to switch into high gear. If it does, we need to be ready for it" (Pasztor 2017, p. 429).

As has been noted, SAI is one of several forms of geoengineering. The most often cited short description of geoengineering is the Royal Society's: "the deliberate large-scale manipulation of the planetary environment to counteract anthropogenic climate change" (The Royal Society 2009, p. 1). There are two general types of geoengineering: carbon dioxide removal (CDR – sometimes referred to as negative emissions technologies or NETS) and solar radiation management (SRM). Stratospheric aerosol injection belongs to the SRM-types of geoengineering.

There are various types of CDR or NETS. Four that receive the greater amount of attention are: bio-energy with carbon capture and storage (BECCS), biochar, ocean fertilization, enhanced weathering, and direct air capture. In BECCS biomass (organic material from plants) is burnt and the CO2 produced is captured and stored. Biochar is the heating of biomass in the absence of oxygen to produce a form of charcoal that does not release its carbon for thousands of years. Fertilizing the ocean causes the rapid growth of organisms (principally plankton), which then die, sink to the ocean floor and sequester the carbon they absorbed. With enhanced weathering "the idea is to control levels of atmospheric CO2 through natural chemical weathering processes that draw CO2 out of the atmosphere (referred to as carbonation) and sequester it in newly-formed rock mineral, magnesium carbonate" (Geoengineering Monitor 2018). Carbon dioxide can also be captured directly from the atmosphere using extractors. Useful by-products can be created from the carbon captured in this process.

Solar radiation management "refers to the deliberate alteration of the earth's radiative balance in order to reduce the risks attributed to the accumulation of greenhouse gases in the atmosphere" (Stavins and Stowe 2019, p. 1). Three principal types of SRM are: marine cloud brightening, cirrus thinning and Stratospheric Aerosol Injection. But any process that prevents solar radiation from reaching the earth's surface or reflects, or helps to reflect, solar radiation away from the earth's surface is a form of SRM. Thus, painting rocks, roofs and roads white, growing crops that reflect solar radiation and using space reflectors can be included under this type of geoengineering. Marine cloud brightening involves "adding cloud condensation nuclei (a specific class of aerosols), such as sea salt, to specific kinds of low-lying clouds over the ocean, with the goal of increasing the reflectivity or lifetime of these clouds" (Keith and Irvine 2019, p. 19). Another form of SRM is to add "ice nuclei (another class of aerosols) to high-altitude cirrus clouds, with the goal of reducing the density of such clouds" (Keith and Irvine 2019, p 20).

Stratospheric Aerosol Injection is the most likely form of SRM to be implemented during the final hundred years of humanity. In part, this is because SAI is less expensive than other forms of SRM. One writer has suggested that SAI "would likely be startlingly *cheap* to implement (in direct cost terms), on the order of tens of billions of dollars per year, while CDR would likely be very *expensive* to deploy at scale, on the order of trillions of dollars" (Horton 2015 emphasis in original).

SAI is also likely to prove effective and rapid. According to Keith and Irvine, "there is high confidence that stratospheric aerosols could achieve sufficient RF [radiative forcing] to off-set half the RF from a doubling of CO2 concentrations..." (2019, p 20). But there are other problems with NETS/CDR that make it less attractive than SAI. Blackstock and Long argued that "technical challenges and large uncertainties surrounding large-scale CDR deployment, along with long delays in the climatic response to carbon forcing, mean that it would take decades to have notable effect", they also claimed that SAI "could substantially influence the climate in months, but with much greater uncertainty about the net effects" (2010, p. 527).

Though it may be cheaper, faster and more effective than CDR, SAI "would entail *high risks* at global scale, in contrast to the *low risks* characteristic of CDR methods (with the exception of ocean fertilization), which would generally involve more familiar technologies with localized direct effects" (Horton 2015 emphasis in original). The risks associated with SAI are collective risks. Even if one actor (the so called 'Greenfinger' - a reference to the Bond villain Goldfinger), government or coalition of governments introduced SAI, all people would be affected.

For an important feature of SAI is that its effects are world-wide and cannot be limited to one country and all share in the risks. But this does not mean that its effects will be the same for all countries (as will be discussed below). This might be deliberate, as "a fairly uniform global aerosol layer could be created, or the aerosol layer could be thicker at high latitudes or in one hemisphere or the other" (Keith and Irvine 2019, p 20). It is more likely to be accidental. This is why one proposal is to inject sulphur dioxide "into the stratosphere at four different latitudes rather than only one in order to better compensate for the pattern of warming due to increased GHG concentrations and reduce the magnitude of residual climate differences" (Hourdequin 2019, p. 1245).

Another important feature of SAI is that the aerosols are likely to have a 1 to 2-year lifespan and will have to be replenished if they are going to maintain their effectiveness. The cost of SAI, then, is an ongoing cost. And, once it is started, it will have to be maintained until the concentration of greenhouse gases in the atmosphere falls to a level at which global warming will cease. The probable consequence of an abrupt cessation of SAI is that the global average surface temperature would rise rapidly. The aerosols are also likely to constitute pollutants, but this might depend on the aerosol and the effects are subject to debate. So the health risks remain unclear.

## *The Politics of Stratospheric Aerosol Injection*

Despite its inherent risks, ongoing costs and differential effects, SAI is highly likely to be introduced during the final hundred years of humanity. SAI does not address species depletion and extinction and will contribute to the increasing toxicity of the planet, however, and fails in a near worst-case scenario. So it is not the answer to catastrophic human-caused environment change and, at best, may simply delay the start of the final hundred years of humanity. SAI also raises many significant questions with respect to its governance. These questions begin with the research and development processes that produce the particular SAI program or programs that will be delivered. Further questions relate to which organizations fund and provide SAI programs, by whom and how it, or they, will be monitored and assessed, regulated and maintained. Compensation for those countries that are deleteriously affected by SAI will also emerge as a question requiring resolution. While the possibility of counter-engineering, with SAI being weaponized, also merits discussion when it comes to the politics of SAI.

This part addresses each of the seven questions with respect to SAI that have a political aspect. The first questions concern research and development. Who conducts the research, how it is conducted and who evaluates it are all important political questions that arise with respect to SAI. Other questions concern the provision of SAI, its monitoring, its regulation and its maintenance (as noted earlier, abrupt cessation of SAI is likely to have disastrous effects). Given that SAI is likely to have a deleterious effect on some countries, questions of compensation also arise. Even if compensation is available and especially when it is not, some of those in countries harmed by SAI might resort to counter-geoengineering. This raises possibilities with respect to SAI being weaponized.

### Research and Development

One problem with respect to research and development in this field is that thinking might narrow too quickly if an appropriate research and development process for SAI is not adopted. Gardiner and Fragnière wrote of the need to pay attention to "the politics of experimentation in geoengineering—the things that are held to be certain, the things regarded as uncertain and worthy of investigation and the things regarded as out-of-bounds (2018, p. 859). In this context, the groupthink that is part of the politics of science (Chap. 2) might work to constrain options. This is one reason that Victor rejected the consensus approach of the IPCC in favour of "multiple competing assessments by small groups linked to active research programmes—reviewed and published openly…" (Victor 2008, p. 325).

A significant concern with respect to SAI is that it will become part of processes that maintain the dominance of countries of the Global North over those of the Global South. As Biermann and Möller put it, "a global discourse and epistemic community is taking shape that might chart the way towards the largest purposeful alteration of planetary systems ever imagined by people— yet this discourse is at present still driven and increasingly legitimized by only few countries in the North" (2019, p. 163). For Carlson and Trisos, "vulnerable populations have not yet been included as equal partners in deliberations, planning, and research concerning SRM" (2018, p. 449).

The skepticism of some of those in the Global South toward the science produced by scientists of the Global North is likely to lead them to be suspicious with respect to the science of SAI. Not including scientists from the Global South in the research and development process would heighten this suspicion. This is why SAI research programs, according to Blackstock and Long, "should include international collaboration, communicate with developing nations, and prioritize research that has global versus national benefits. These steps will limit the new problems geoengineering research heaps on an already strained global climate agenda, preserving options for future international cooperation" (2010, p. 527).

## Provision

The first question with respect to the deployment of SAI concerns who will provide it. Four general scenarios can, and have, been outlined. One is of an international body overseeing an international system for research, provision and monitoring and assessment. Another is of system overseen by a regional body, or a group comprised of representatives from specific countries. Yet another scenario for the provision of SAI is by an individual government. The final scenario is that of an individual who has the resources to fund research into and the provision of SAI (often referred to, due to Victor (2008), as a "Greenfinger").

A significant fear amongst many authors who have written about SAI is that a government or person might unilaterally decide to implement SAI. While others raise doubts as to whether a lone nation or individual would do this, given the possibility of retaliation from the international community. But this is a discussion of the politics of desperate measures and the possibility of what Weitzman called a "*free-driver*" (Weitzman 2015) remains real. For, as conditions deteriorate more rapidly in a particular region, the people of that region will look to their government to provide SAI, and some leaders might seek to boost their support through unilateral deployment (Victor 2019, p. 43).

## Monitoring and Assessment

The question of how SAI will be monitored and assessed is an important one, as it can predetermine the approach taken to addressing catastrophic human-caused environment change. A consequence of the IPCC's decision to frame 'climate change' in terms of projections with respect to average global surface temperature was to enable SAI itself. Because SAI addresses average global surface temperature it comes to be a 'solution' to the problem of 'global warming' – which is, by definition, temperature related – even though it does not address other causes of catastrophic human-caused environment change (which means that greenhouse gasses may still accumulate in the atmosphere).

An alternative means for monitoring and assessing 'climate change' is through measuring the concentration of greenhouse gases in the atmosphere. While this measure is likely to have its own problems – such as with respect to adequate data collection – it does not focus attention on temperature and cannot be addressed via SAI. However, as Ricke wrote, "global temperature targets are … a largely entrenched element of how … climate-research communities approach global climate governance." And the fact that researchers "cannot unilaterally veto the use of global temperature targets in climate policy goal setting, [means that] identifying a way to reconcile the outcomes of [SAI] deployment with the temperature-target standard must be a priority" (Ricke 2019, p. 97).

Monitoring and assessing the effects of SAI will be crucial features of the system for governing its implementation and continuing provision. Its very "deployment strongly depends on states' ability to monitor each other's deployment" (Philippe 2019, p. 71). This presupposes a capacity to observe SAI's effects in a timely manner. And "the consequences of geoengineering – intended as well as unintended – could be very difficult to predict" (Humphreys 2011, p. 116). Measuring global average temperature is not enough, Philippe argued, "because it would take years to detect the effects of [SAI] with high confidence." For him, "new *in-situ*, ground-based, and possibly space-based observation and measurement assets would be required in the years preceding deployment" (Philippe 2019, p. 71). This will delay and increase the cost of implementing SAI.

But even if SAI can be monitored effectively and technologies and techniques for monitoring SAI's effects are introduced, the politics of the final hundred years of humanity is, in part, a politics of mistrust on the parts of those in countries of the Global South toward those of the Global North. One source of this mistrust, as was discussed in Chap. 3, is that scientists of the Global North dominate climate science and have not include scientists from the Global South in all of their research programs. As Philippe pointed out, "an important attribute for transnational information acquisition and sharing systems is the ability to demonstrate that the data collected by various sensors or sources are genuine and can be trusted by all participating states" (2019, p. 71).

## Regulation

The most important political question with respect to the use of SAI to counteract global warming is how it will be governed. For no matter who and how it is provided and monitored and assessed, support for SAI will be greatly affected by, if it does not depend upon, how an SAI program is regulated. This raises a variety of questions concerning the nature of the organization or organizations that oversee SAI's development, implementation and monitoring and assessment. These questions relate to its, or their, composition, funding, manner of decision-making, accountability and maintenance. The latter is of particular concern, as "any large-scale use of solar engineering would likely condemn future generations to continue with it for hundreds of years" (Pasztor 2017, p 423).

Some scholars have argued that a de facto system of governance is already in place. Talberg and colleagues, for example, claimed to have shown "that geoengineering is loosely governed in an unplanned manner by a mix of international agreements designed for other purposes" (Talberg et al. 2018, p. 247). While Gupta and Möller rejected the conventional view and argued that geoengineering "is already a governed space." After acknowledging "the relative absence of formal governance arrangements," they asserted "that an emerging de facto governance of CE is discernible, by which we understand, following Rip, sources of governance that are unacknowledged and unrecognized as seeking to govern, even as they exercise governance effects" (Gupta and Möller 2019, p. 481).

The vast majority of those writing about geoengineering in general, and SAI in particular, argue that it is not currently regulated, and that no system is emerging for regulating most forms of geoengineering. Some see developing such a system as an urgent priority; though Victor has cautioned that "the more that western countries insist that governance be in place before deployment occurs, the higher the odds that deployment will occur without a system of governance to deal with the consequences" (Victor 2019, p. 41). Yet, according to Nye, given the time it took for international cooperation to emerge with respect to cyber security and nuclear weapons, when it comes to geoengineering, "we should think in terms of decades…" (Nye 2019, p. 56). Those in positions of political power will have a strong interest in regulating the behaviour of other leaders, though, so, according to Barrett, "states would be willing and perhaps even eager to negotiate an agreement involving mutual restraint" (2019, pp. 34–5).

The form that the organization or organizations that regulate SAI will take may depend on the understanding of the climate emergency to which SAI is a response. If the emergency is understood to be a global problem, then "governance is needed for purposes such as agreeing that geoengineering should be launched, deciding how to share costs, and managing possible side-effects" (Victor 2019, p. 42). Some authors have suggested that the UN Security Council is "the only existing institution with relevant, binding decision-making authority …, but it would not be able to limit solar geoengineering by the permanent five member states, which have veto power" (Bodansky 2019, p. 119). But the general principle with respect to regulation at this level is that "only limited international governance functions are likely to prove acceptable to states" (Bodansky 2019, p. 122).

If the climate emergency is regional, then to the list of jobs for governing a global emergency must be added "the extra task of managing conflicts between countries that wish to deploy geoengineering and those who oppose action…" (Victor 2019, p. 42). If the emergency is local or is perceived to be more pressing by the people of one locality, then different questions may need to be resolved. Thus, "whereas a deployment response to a global climate emergency may require cost sharing and the implementation of complementary projects, responding to an imagined local climate emergency may instead require dispute resolution mechanisms and risk assessment" (Jinnah 2019, p. 143).

Should questions associated with what type of organization (or organizations) regulates the provision of SAI be resolved, the next questions concern the principles through which it ought to be regulated. The Oxford Principles represent an early attempt to develop governing principles for geoengineering that could be applied to SAI. These are:

Principle 1: Geoengineering to be regulated as a public good. …
Principle 2: Public participation in geoengineering decision-making. …
Principle 3: Disclosure of geoengineering research and open publication of results. …
Principle 4: Independent assessment of impacts. …
Principle 5: Governance before deployment. … (Rayner et al. 2013, p. 502–3)

Of other proposals, Stilgoe has suggested the following governing principles: "(1) comparative benefit; (2) accountability; (3) transparency; (4) substantive justice; and (5) procedural justice" (2016, p. 331).

## Maintenance

As has already been noted, because it does not address the concentration of greenhouse gases in the atmosphere SAI will have to be maintained until the concentration of those gases falls to a level at which global warming ceases (or, perhaps, slows to the point of ceasing to have catastrophic effects). The introduction of SAI, then, will require "long-term institutional guarantees against premature termination" (Pasztor 2019, p. 149). Stock expressed the common view that "even a temporary suspension of [SAI] would result in very rapid and large changes in temperature with highly uncertain climate consequences" (Stock 2019, p. 104). While, as Honegger pointed out with respect to the 'termination shock" induced by the premature cessation of SAI: "the likelihood of termination shock rests heavily on the stability and effectiveness (or lack thereof) of … governance arrangements" (2019, p. 78). Thus, the provision of SAI will require maintaining a sustained effort "over the very long time-frames required – longer than the lifetimes of most human institutions and *much* longer than the terms in office of key decision-makers" (Bunn 2019, p. 51).

## Compensation

The effects of SAI are likely to be different in different regions. While "studies have shown that the climate response to geoengineering may be roughly linear in terms of global mean temperature, … regional and seasonal changes and hydrologic responses exhibit nonlinearities" (Boettcher et al. 2019, p. 246). According to Keith and Irvine, "there is no way to deploy solar geoengineering that would be regarded as optimal by all actors in all regions. The effects of SG [solar geoengineering] would be different in different regions, different regions would be exposed to different risks …" (2019, p. 23). A commonly cited example of the differential effect of SAI is that it "may disrupt hydrological cycles, precipitating famine in African and Asian countries that grow monsoon crops" (Bas and Mahajan 2019, p. 67). So, if it is not to be a continuation of a process in which those in more powerful countries simply (studiously) ignore the effects of their actions on those in less powerful countries, an SAI program must include a mechanism for effectively resolving "issues around liability and compensation in case of unequal outcomes …" (Pasztor 2019, p. 149).

### Counter-Geoengineering – Weaponized SAI

Another question that arises with respect to the politics of SAI concerns how to address the possibility of counter-geoengineering, that is "the use of technical means to negate the change in radiative forcing caused by SRM deployment" (Parker et al. 2018, p. 1058). In the case of SAI, such counter-geoengineering could involve rendering SAI "particles inert by, for example, injecting a base to counteract the sulphate aerosols most commonly considered for [SAI]." Another method is to release "warming agents such as greenhouse gases or specially engineered solid particles to counter the change in radiative forcing caused by [SAI]. Both forms of CG are possible, but neither currently exists …" (MacMartin et al. 2019, p. 154).

As was noted in the previous discussion, SAI may have profoundly different effects on different regions and "one can envisage a situation where some countries actively seek out solar engineering's predicted regional effects, despite the concerns of others" (Pasztor 2017, p. 423). Bas and Mahajan were worried that geoengineering could "produce a 'tug-of-war' over the climate" (2019). As they wrote, and was discussed in Chap. 6, "in some cases, warmer temperatures may confer geopolitical benefits. As the Arctic melts, proximate countries such as Russia may gain access to oil and natural gas reserves and strategically important shipping routes …" (Bas and Mahajan 2019, p. 49).

Unfortunately, concerns with respect to counter-geoengineering have escalated, for some, to concerns with respect to the weaponizing of counter-geoengineering. Stock has warned that "finely tuned [SAI] could be weaponized, with potentially large ecosystem damages as well as human costs" (2019, p. 104). While a contribution to *The Bulletin of the Atomic Scientists*, warned that, "already, military organizations are taking an interest. The Pentagon's Defense Advanced Research Projects Agency convened a meeting in 2009 to consider geoengineering" (The Bulletin of the Atomic Scientists 2014, p. 22).

## Conclusion

While there are good reasons not to discuss geoengineering and therefore to remain silent with respect to SAI, the time has passed at which such restraint makes sense. So, the politics of SAI must be discussed. The politics of SAI will be either simple or complex. It will be simple if the leaders of one or a small coalition of countries decide to impose SAI and ignore opposition from those in other countries. It will 'simply' be a form of warfare, with those who do not want it either engaging in counter-geoengineering or seeking to disrupt the SAI that has been imposed on them. There will be no need to consider questions associated with research and development, provision, monitoring and assessment, regulation and compensation. If SAI is undertaken at an international level, then all these questions arise, and the process of introducing and maintaining SAI becomes complex.

Other forms of geoengineering are available, CDR is slow and expensive and the other forms of SRM will not have the impact, for better or for worse, than SAI will have. Even if it is introduced, works from the outset and does not go wrong as it continues, SAI will not prevent the onset of the final hundred years of humanity. For "the delayed implementation of solar geoengineering assumes ecosystems survive long enough to see the potential benefits" (Boettcher et al. 2019, p. 247).

Radical population reduction and SAI will not be the only issues with respect to which the politics of desperate measures has to be understood. The various forms of escape that are attempted by individuals, groups and government will raise their own political questions. But radical population reduction and SAI represent two of the likeliest responses during the final hundred years of humanity and the two with the greatest potential for promoting conflict during those final hundred years. With foresight, careful planning and goodwill on all parts, the conflicts may not prove divisive and add to the other conflicts that will occur during the final hundred years of humanity. But foresight, careful planning and goodwill on all parts might have resulted in catastrophic human-caused conflict being addressed and a trajectory toward the final hundred years of humanity averted. The Global North-Global South divide will constitute a crucial context within which the politics of the desperate measures radical population reduction and SAI is enacted. In both cases, the tensions between countries of the Global North and those of the Global South are likely to intensify.

## References


Abbey, E. (1975). *The monkey wrench gang*. Philadelphia: Lippincott Williams & Wilkins.

Ahmed, N. (2011, December 23). Development of the soul. *Dawn*. https://www.dawn.com/news/682589/development-of-the-soul. Accessed 10 July 2019

Ataullahjan, A., Mumtaz, Z., & Vallianatos, H. (2019). Family planning, Islam and sin: Understandings of moral actions in Khyber Pakhtunkhwa, Pakistan. *Social Science & Medicine*, *230*, 49–56. https://doi.org/10.1016/j.socscimed.2019.03.011.

Barrett, S. (2019). Some thoughts on Solar Geoengineering Governance. In R. N. Stavins and R. C. Stowe (Eds.) *Governance of the deployment of solar geoengineering*. Cambridge, Mass.: Harvard Project on Climate Agreements. https://www.belfercenter.org/sites/default/files/files/publication/harvard_project_sg_governance-briefs_volume_feb_2019.pdf. Accessed 16 July 2019.

Bas, M. & Mahajan, A. (2019). Generating and Managing International Conflict How Geoengineering can Produce a "Tug-of-War" over the Climate. In R. N. Stavins and R. C. Stowe (Eds.) *Governance of the deployment of solar geoengineering*. Cambridge, Mass.:

Harvard Project on Climate Agreements. https://www.belfercenter.org/sites/default/files/files/publication/harvard_project_sg_governance-briefs_volume_feb_2019.pdf. Accessed 16 July 2019.

Batchelor, D. A. F. (2019). Islamic perspectives on curbing population growth to promote earth's sustainability. *Islam and Civilisational Renewal (ICR)*, *10*(1), 21–46. https://www.icrjournal.org/icr/index.php/icr/article/view/769.

Bendix, D. (2016). From fighting underpopulation to curbing population growth: Tracing colonial power in German development interventions in Tanzania. *Postcolonial Studies*, *19*(1), 53–70. https://doi.org/10.1080/13688790.2016.1228137.

Biermann, F., & Möller, I. (2019). Rich man's solution? Climate engineering discourses and the marginalization of the Global South. *International Environmental Agreements: Politics, Law and Economics*, 19, 151–167. https://doi.org/10.1007/s10784-019-09431-0.

Blackstock, J. J., & Long, J. C. (2010). The politics of geoengineering. *Science*, *327*(5965), 527–527. https://doi.org/10.1126/science.1183877.

Bodansky, D. (2019). Solar geoengineering and international law. In R. N. Stavins & R. C. Stowe (Eds.) *Governance of the deployment of solar geoengineering*. Cambridge, Mass.: Harvard Project on Climate Agreements. https://www.belfercenter.org/sites/default/files/files/publication/harvard_project_sg_governancebriefs_volume_feb_2019.pdf. Accessed 16 July 2019.

Boettcher, M., Chai, F., Cullen, J., Goeschl, T., Lampitt, R., Lenton, A., … & Wanninkhof, R. (2019). High Level Review of a Wide Range of Proposed Marine Geoengineering Techniques. http://oceanrep.geomar.de/46221/1/rs98e.pdf. Accessed 14 July 2019.

Bowen, J., & Purkis, J. (2018). Introduction: Why anarchism still matters. In J. Purkis & J. Bowen (Eds.) *Changing anarchism*. Manchester University Press.

Bradshaw, C. J., & Brook, B. W. (2014). Human population reduction is not a quick fix for environmental problems. *Proceedings of the National Academy of Sciences*, *111*(46), 16610–16615. https://doi.org/10.1073/pnas.1410465111.

Bratton, S. (1992). Six billion and more: Human population regulation and Christian ethics. Westminster: John Knox Press.

Bunn, M. (2019). Governance of Solar Geoengineering: Learning from Nuclear Regimes. In R. N. Stavins and R. C. Stowe (Eds.) *Governance of the deployment of solar geoengineering*. Cambridge, Mass.: Harvard Project on Climate Agreements. https://www.belfercenter.org/sites/default/files/files/publication/harvard_project_sg_governance-briefs_volume_feb_2019.pdf. Accessed 16 July 2019.

Campbell-Arvai, V., Hart, P. S., Raimi, K. T., & Wolske, K. S. (2017). The influence of learning about carbon dioxide removal (CDR) on support for mitigation policies. *Climatic Change*, *143*(3–4), 321–336. https://doi.org/10.1007/s10584-017-2005-1.

Carlson, C. J., & Trisos, C. H. (2018). Climate engineering needs a clean bill of health. *Nature Climate Change*, *8*(10), 843. https://doi.org/10.1038/s41558-018-0294-7.

Corry, O. (2017). The international politics of geoengineering: The feasibility of Plan B for tackling climate change. *Security Dialogue*, *48*(4), 297–315. https://doi.org/10.1177/0967010617704142.

Djupe, P. A., Neiheisel, J. R., & Conger, K. H. (2018). Are the politics of the Christian right linked to state rates of the nonreligious? The importance of salient controversy. *Political Research Quarterly*, *71*(4), 910–922.

Elchardus, M., & Spruyt, B. (2016). Populism, persistent republicanism and declinism: An empirical analysis of populism as a thin ideology. *Government and Opposition*, *51*(1), 111–133. https://doi.org/10.1017/gov.2014.27.

Ferrari, S. (1988). Separation of church and state in contemporary European society, *Journal of Church and State* 30(3), 533–547. https://doi.org/10.1093/jcs/30.3.533.

Fletcher, R. (2018). License to kill: Contesting the legitimacy of green violence. *Conservation and Society*, 16(2), 147–156. https://doi.org/10.4103/cs.cs_16_148.

Frazer, E., & Hutchings, K. (2019). Anarchist ambivalence: Politics and violence in the thought of Bakunin, Tolstoy and Kropotkin. *European Journal of Political Theory*, *18*(2), 259–280. https://doi.org/10.1177/1474885116634087.

Gardiner, S. M., & Fragnière, A. (2018). Geoengineering, Political Legitimacy and Justice. https://doi.org/10.1080/21550085.2018.1562524.

Geoengineering Monitor. (2018, May 9), Enhanced Weathering (Technology Factsheet) http://www.geoengineeringmonitor.org/2018/05/enhanced-weathering-factsheet/. Accessed 19 July 2019.

Gibbs, N. (2019). Modern Constitutional Legitimacy and Political Theology: Schmitt, Peterson and Blumenberg. *Law and Critique*, *30*(1), 67–89. https://doi.org/10.1007/s10978-018-9228-9

Gökarıksel, B., Neubert, C., & Smith, S. (2019). Demographic fever dreams: Fragile masculinity and population politics in the rise of the global right. *Signs: Journal of Women in Culture and Society*, *44*(3), 561–587. https://doi.org/10.1086/701154.

Goodkind, D. (2017). The astonishing population averted by China's birth restrictions: Estimates, nightmares, and reprogrammed ambitions. *Demography*, *54*(4), 1375–1400. https://doi.org/10.1007/s13524-017-0595-x.

Gray, P. W. (2007). Political theology and the theology of politics: Carl Schmitt and Medieval Christian political thought. Humanitas, 20(1–2), 175–200.

Greenhalgh, S. (2018). Making demography astonishing: lessons in the politics of population science. *Demography*, *55*(2), 721–731. https://doi.org/10.1007/s13524-018-0660-0.

Gunderson, R., Stuart, D., & Petersen, B. (2018). The political economy of geoengineering as plan B: Technological rationality, moral hazard, and new technology. *New Political Economy*, 1–20. https://doi.org/10.1080/13563467.2018.1501356.

Gupta, A., & Möller, I. (2019). De facto governance: how authoritative assessments construct climate engineering as an object of governance. *Environmental Politics*, *28*(3), 480–501. https://doi.org/10.1080/09644016.2018.1452373.

Hartmann, B., Hendrixson, A., & Sasser, J. (2016). Population, sustainable development and gender equality. In M. Leach (Ed.). *Gender Equality and Sustainable Development* (Vol. 56, No. 81, pp. 56–81). London: Routledge in association with GSE Research.

Hendrixson, A. (2019). Population control in the troubled present: The '120 by 20' target and implant access program. *Development and Change*, *50*(3), 786–804. https://doi.org/10.1111/dech.12423.

Hesketh, T., Lu, L., & Xing, Z. W. (2005). The effect of China's one-child family policy after 25 years. N Engl J Med, 353,1171–1176. https://doi.org/10.1056/NEJMhpr051833.

Honegger, M. (2019). Assessing Solar Geoengineering – What, Who, and How? In R. N. Stavins and R. C. Stowe (Eds.) *Governance of the deployment of solar geoengineering*. Cambridge, Mass.: Harvard Project on Climate Agreements. https://www.belfercenter.org/sites/default/files/files/publication/harvard_project_sg_governance-briefs_volume_feb_2019.pdf. Accessed 16 July 2019.

Horton, J. (2019). Evaluating Solar Geoengineering Deployment Scenarios. In R. N. Stavins and R. C. Stowe (Eds.) *Governance of the deployment of solar geoengineering*. Cambridge,

Horton, J. B. (2015, Dec. 16). Why We Should Treat SRM and CDR Separately *Forum for Climate Engineering Assessment*. http://ceassessment.org/why-we-should-treat-srm-and-cdr-separately-joshua-b-horton/. Accessed 18 July 2019.

Hourdequin, M. (2019). Geoengineering justice: the role of recognition. *Science, Technology, & Human Values*, *44*(3), 448–477. https://doi.org/10.1177/0162243918802893.

Humphreys, D. (2011). Smoke and mirrors: Some reflections on the science and politics of geoengineering. *The Journal of Environment & Development*, *20*(2), 99–120. https://doi.org/10.1177/1070496511405302.

Jinnah, S. (2019). Building a Governance Foundation for Solar Geoengineering Deployment. In R. N. Stavins and R. C. Stowe (Eds.) *Governance of the deployment of solar geoengineering*. Cambridge, Mass.: Harvard Project on Climate Agreements. https://www.belfercenter.org/sites/default/files/files/publication/harvard_project_sg_governance-briefs_volume_feb_2019.pdf. Accessed 16 July 2019.

Keith, D. & Irvine, P. (2019). The Science and Technology of Solar Geoengineering: A Compact Summary Introduction. In R. N. Stavins and R. C. Stowe (Eds.) *Governance of the deployment of solar geoengineering*. Cambridge, Mass.: Harvard Project on Climate Agreements. https://www.belfercenter.org/sites/default/files/files/publication/harvard_project_sg_governance-briefs_volume_feb_2019.pdf. Accessed 16 July 2019.

King James Bible Online. https://www.kingjamesbibleonline.org/Genesis-1-28/. Accessed 7 July 2019.

Kinna, R., Prichard, A., & Swann, T. (2019). Occupy and the constitution of anarchy. *Global Constitutionalism*, *8*(2), 357–390. https://doi.org/10.1017/S204538171900008X.

Korkut, U., & Eslen-Ziya, H. (2016). The Discursive Governance of Population Politics: The Evolution of a Pro-birth Regime in Turkey. *Social Politics: International Studies in Gender, State & Society*, *23*(4), 555–575. https://doi.org/10.1093/sp/jxw003.

LeVasseur, T. (2017). Decisive ecological warfare: Triggering industrial collapse via deep green resistance. *Journal for the Study of Religion, Nature & Culture*, *11*(1), 109–130. https://doi.org/10.1558/jsrnc.29799.

Lin, A. (2019). The Relevance of the Climate Change Regime to Governance of Solar Geoengineering. In R. N. Stavins and R. C. Stowe (Eds.) *Governance of the deployment of solar geoengineering*. Cambridge, Mass.: Harvard Project on Climate Agreements. https://www.belfercenter.org/sites/default/files/files/publication/harvard_project_sg_governance-briefs_volume_feb_2019.pdf. Accessed 16 July 2019.

Ma, J. (2013, 6 May). China's barbaric one-child policy. *The Guardian*, https://www.theguardian.com/books/2013/may/06/chinas-barbaric-one-child-policy. Accessed 15 July 2019.

MacMartin, D. G., Wang, W., Kravitz, B., Tilmes, S., Richter, J. H., & Mills, M. J. (2019). Timescale for detecting the climate response to stratospheric aerosol geoengineering. *Journal of Geophysical Research: Atmospheres*, *124*(3), 1233–1247. https://doi.org/10.1029/2018JD028906.

Malthus, T. (1826). *An essay on the principle of population*. 6th edition. London: John Murray. https://www.econlib.org/library/Malthus/malPlong.html?chapter_num=47#book-reader. Accessed 7 Jan 2020.

Manes, C. (1990). *Green rage: Radical environmentalism and the unmaking of civilization*. New York: Little, Brown & Co.

Merk, C. (2019). Moral Hazard and Solar Geoengineering. In R. N. Stavins and R. C. Stowe (Eds.) *Governance of the deployment of solar geoengineering*. Cambridge, Mass.: Harvard Project on Climate Agreements. https://www.belfercenter.org/sites/default/files/files/publication/harvard_project_sg_governance-briefs_volume_feb_2019.pdf. Accessed 16 July 2019.

Nye, J. S. (2019). Notes on Insights from Other Regimes: Cyber. In R. N. Stavins and R. C. Stowe (Eds.) *Governance of the deployment of solar geoengineering*. Cambridge, Mass.: Harvard Project on Climate Agreements. https://www.belfercenter.org/sites/default/files/files/publication/harvard_project_sg_governance-briefs_volume_feb_2019.pdf. Accessed 16 July 2019.

Ojeda, D., Sasser, J. S., & Lunstrum, E. (2019). Malthus's specter and the anthropocene. *Gender, Place & Culture*, 1–17. https://doi.org/10.1080/0966369X.2018.1553858.

Parker, A., Horton, J. B., & Keith, D. W. (2018). Stopping Solar Geoengineering Through Technical Means: A Preliminary Assessment of Counter-Geoengineering. *Earth's Future*, *6*(8), 1058–1065. https://doi.org/10.1029/2018EF000864.

Pasztor, J. (2017). The need for governance of climate geoengineering. *Ethics & International Affairs*, *31*(4), 419–430. https://doi.org/10.1017/S0892679417000405.

Pasztor, J. (2019). The Road to Solar Geoengineering Governance. In R. N. Stavins and R. C. Stowe (Eds.) *Governance of the deployment of solar geoengineering*. Cambridge, Mass.: Harvard Project on Climate Agreements. https://www.belfercenter.org/sites/default/files/files/publication/harvard_project_sg_governance-briefs_volume_feb_2019.pdf. Accessed 16 July 2019.

Pew Research Center (2013, April 30). The world's Muslims: religion, politics and society. https://www.pewforum.org/2013/04/30/the-worlds-muslims-religion-politics-society-overview/. Accessed 15 July 2019.

Philippe, S. (2019). Monitoring and Verifying the Deployment of Solar Geoengineering. In R. N. Stavins and R. C. Stowe (Eds.) *Governance of the deployment of solar geoengineering*. Cambridge, Mass.: Harvard Project on Climate Agreements. https://www.belfercenter.org/sites/default/files/files/publication/harvard_project_sg_governance-briefs_volume_feb_2019.pdf. Accessed 16 July 2019.

Rayner, S., Heyward, C., Kruger, T., Pidgeon, N., Redgwell, C., & Savulescu, J. (2013). The oxford principles. Climatic Change, 121(3), 499–512. https://doi.org/10.1007/s10584-012-0675-2.

Ricke, K. (2019). The Challenge of Reconciling Global Temperature Targets with the Prospect of Solar Geoengineering. In R. N. Stavins and R. C. Stowe (Eds.) *Governance of the deployment of solar geoengineering*. Cambridge, Mass.: Harvard Project on Climate Agreements. https://www.belfercenter.org/sites/default/files/files/publication/harvard_project_sg_governance-briefs_volume_feb_2019.pdf. Accessed 16 July 2019.

Ritchie, H., & Roser, M. (2019) $CO_2$ and other Greenhouse Gas Emissions. *OurWorldInData.org*. https://ourworldindata.org/co2-and-other-greenhouse-gas-emissions. Accessed 3 July 2019.

Ruthven, M. (2016). Mosque and state: The future of political Islam. *Foreign Affairs*, *95*, 142.

Stavins, R. N. & Stowe, R.C. (2019). Introduction. In R. N. Stavins and R. C. Stowe (Eds.) *Governance of the deployment of solar geoengineering*. Cambridge, Mass.: Harvard Project on Climate Agreements. https://www.belfercenter.org/sites/default/files/files/publication/harvard_project_sg_governance-briefs_volume_feb_2019.pdf. Accessed 16 July 2019.

Stilgoe, J. (2016). Geoengineering as collective experimentation. *Science and Engineering Ethics*, *22*(3), 851–869. https://doi.org/10.1007/s11948-015-9646-0.

Stock, J. H. (2019). An Economist's View on Solar Geoengineering Governance under Uncertainty. In R. N. Stavins and R. C. Stowe (Eds.) *Governance of the deployment of solar geoengineering*. Cambridge, Mass.: Harvard Project on Climate Agreements. https://www.belfercenter.org/sites/default/files/files/publication/harvard_project_sg_governance-briefs_volume_feb_2019.pdf. Accessed 16 July 2019.

Talberg, A., Christoff, P., Thomas, S., & Karoly, D. (2018). Geoengineering governance-by-default: an earth system governance perspective. *International Environmental Agreements: Politics, Law and Economics*, *18*(2), 229–253. https://doi.org/10.1007/s10784-017-9374-9.

The Bulletin of the Atomic Scientists 2014, Vol. 70(3) 17–26. sagepub.co.uk/journalsPermissions.nav. https://doi.org/10.1177/0096340214531173.

The Royal Society (2009). *Geoengineering the climate: science, governance and uncertainty* RS Policy document 10/09. RS1636. https://royalsociety.org/topics-policy/publications/2009/geoengineering-climate/. Accessed 19 July 2019.

Thiele, L. P. (2019). Geoengineering and sustainability. *Environmental Politics*, *28*(3), 460–479. https://doi.org/10.1080/09644016.2018.1449602.

Tylecote, A. (2019). Biotechnology as a new techno-economic paradigm that will help drive the world economy and mitigate climate change. *Research Policy*, *48*(4), 858–868. https://doi.org/10.1016/j.respol.2018.10.001.

Victor, D. G. (2008). On the regulation of geoengineering. *Oxford Review of Economic Policy*, *24*(2), 322–336. https://doi.org/10.1093/oxrep/grn018.

Victor, D. G. (2019). Governing the Deployment of Geoengineering: Institutions, Preparedness, and the Problem of Rogue Actors. In R. N. Stavins and R. C. Stowe (Eds.) *Governance of the deployment of solar geoengineering*. Cambridge, Mass.: Harvard Project on Climate Agreements. https://www.belfercenter.org/sites/default/files/files/publication/harvard_project_sg_governance-briefs_volume_feb_2019.pdf. Accessed 16 July 2019.

Wang, X., & Zhang, J. (2018). Beyond the Quantity–Quality tradeoff: Population control policy and human capital investment. *Journal of Development Economics*, *135*, 222–234. https://doi.org/10.1016/j.jdeveco.2018.04.007.

Weitzman, M. L. (2015). A voting architecture for the governance of free-driver externalities, with application to geoengineering. *The Scandinavian Journal of Economics*, *117*(4), 1049–1068. https://doi.org/10.1111/sjoe.12120.

Wilde, M. J., & Hopkins, K. (2018). From eugenicists to family planners: America's religious promoters of contraception. In Z. Amarin (Ed.), *Family Planning*, 13–32. https://doi.org/10.5772/intechopen.72378.

# Conclusion

The politics of catastrophic human-caused climate change leading to the beginning of the final hundred years of humanity has international, national and local dimensions. The politics that operates through these dimensions will be conducted in light of matters raised in Chaps. 2, 3 and 4. For, scientists will continue to struggle to assert their authority due to the loss of status they suffered when their claims to possess absolute Truth were exposed by climate change deniers. This will become particularly important with respect to the introduction of SAI. Just as those in governments will continue to undertake their role in preserving the economic, social and political order in which they achieved their power. The various factors, discussed in Chap. 3, that have served to keep humanity on its path toward the beginning of its final hundred years will continue to function. People may lose faith in their political and economic order and begin to question those orders. But those with a stake in the system will continue to fund the climate change denial industry that has worked assiduously to ensure that nothing was done, and for some time to come will be done, to address the effects of catastrophic human-caused environment change.

The politics of the final hundred years of humanity's international dimension has three sub-dimensions. The first of these includes war, resource conflict and border conflict. To engage in any of these forms of conflict requires a military capacity that only a limited number of countries currently possess, however, and even fewer will have as the effects of catastrophic human-caused environment change worsen. A second sub-dimension relates to the implementation of international programs to achieve radical population reduction. The third international sub-dimension concerns the introduction of stratospheric aerosol injection. While it will become part of the other dimension, should it become weaponized, SAI raises a variety of questions of governance that must be addressed if the introduction of SAI is not to descend into outright conflict and/or failure.

The national dimension of the politics of the final hundred years of humanity also has two sub-dimensions, but both are contingent on the capacity of the government concerned. Those in governments around the world will be required to address



I. Cook, *The Politics of the Final Hundred Years of Humanity (2030-2130)*,
https://doi.org/10.1007/978-981-15-1259-9

the issues that arise with respect to national adjustment due to catastrophic human-caused environment change – especially desertification and rising sea levels. Some will fail the test and their people will become part of the ever-growing number of people displaced because of catastrophic human-caused environment change.

The other sub-dimension involves radical reduction of the national population and the negotiation of the moral and ethical issues that arise with respect to radical population reduction. This negotiation is likely to return an intensified form of the familiar contest between church and state. Specific national governments will have a capacity to invest in a variety of geoengineering projects, and there remains a possibility that some governments will act on their own to engage in geoengineering. So, a national politics of SAI is possible, though it is likely to escalate to an international level.

The local dimension of the politics of the final hundred years of humanity will be that of providing, where possible, for a community that is being affected by rising sea levels and desertification (which includes ocean deserts caused by acidification). Once again, the crucial question will concern a government's capacity to itself survive the effects of catastrophic human-caused environment change before expending the necessary resources to protect those populations for whom remaining where they are is deemed necessary and facilitating relocation of those populations leaving voluntarily or being forced to leave.

Given what was written about the hubris of climate change deniers, who claim to know that climate scientists are perpetrating a fraud, nothing in this book can be thought to represent 'The Truth'. Thus, the possibility remains that climate change deniers are right, after all, and the greatest fraud in human history will have been committed by literally thousands of scientists who, in one way or another, have contributed to perpetration of that fraud. Should that turn out to be the case, the politics of 'The Great Climate Change Fraud' will become of particular interest. For a considerable challenge will lie in explaining how this fraud was perpetrated despite climate scientists' limited connections and the operation of scientific method, peer review and the other processes that have previously been assumed to result in as close an approximation to the Truth as can be developed through science.

It is traditional, in a work such as this, to conclude with a 'call to arms' involving an exhortation for 'us' to heed the warnings of climate scientists and take measures to address the causes of catastrophic human-caused environment change. As part of such a conclusion, readers would be told that there is still time, and that concerted action, especially by those in the Global North, can prevent catastrophic human-caused climate change and avert a trajectory toward the final hundred years of humanity. This might be conducted by way of an appeal to some inherent 'humanity' or spiritual connection to nature or even simple rationality.

This will not happen here. For the reasons outlined in this book, too little has been done and nothing significantly more will be done to address the causes of catastrophic human-caused environment change and divert from a path toward the final hundred years of humanity. In short, no answer will be provided, when it comes to averting a trajectory toward the final hundred years of humanity, to what can be thought of as THE question of politics (and, for some, as Lenin's question): 'What is to be Done?' about catastrophic human-caused environment change and the onset of the final hundred years of humanity. But crucial political questions will still have to be answered and some answer produced for the question "What is to be done given the failure to address the causes of catastrophic human-caused environment change and the maintenance of a trajectory toward the final hundred years of humanity?" And there are many things to be done and politics is part of almost, if not all, of them.

The manufacturer's authorised representative in the EU is Springer Nature Customer Service Centre GmbH, Europaplatz 3, 69115 Heidelberg, Germany. If you have any concerns regarding our products, please contact ProductSafety@springernature.com

Printed and bound by CPI Group (UK) Ltd, Croydon, CR0 4YY
30/09/2026
02225234-0004